Philippe Bauduin

Wars & Discoveries

Translated by T. Brian Greenhalgh & Michael Greenhalgh, M. Phil. BA (Hons).

CONTENTS

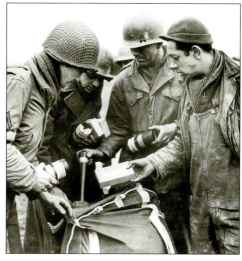

Foreword

The great "disasters" at the beginning of the 20th Century

Three notable events early in the 20th Century mobilised the inventive genius
of mankind :
- The sinking of the Titanic that directly effected the discovery of Sonar then
  Radar.
- The Great War in which great progress was made in care for the wounded
  and more particularly in aviation.
- The widespread epidemic of Spanish Flu in the winter of 1918-1919 which claiming
  more victims than the war itself, led to the discovery later of penicillin and antibiotics.
So it was in the Thirties numerous commodities, appliances and devices taken
for granted today were discovered : antibiotics, sulphonamides, radar, turbo-jets,
synthetic rubber, mineral oils, radiotelephony, and artificial intelligence… but only
the outbreak of the Second World War could give rise to their development
and industrial use.

The Second World War and the Battle of Normandy

In four years of war, nowhere did the belligerents use rival technological advance
more than in Normandy. Three million soldiers armed with the very latest devices :
hundreds of sets of radar and as many V1 ramps and V2 launchers, the first jet aircraft,
portable telephones, computers, logistics of both blood and water hitherto unknown,
dozens of landing-strips, thousands of nylon parachutes and incredible
prefabricated ports (half a century later Arromanches boasts imposing remnants),
confronted the bemused and horrified eyes of the people of Normandy. In the summer
of 1944, superimposed on a Normandy in ruins was a Detroit of vehicles in assembly
lines under the apple trees and an Emirate of pipelines. In addition to which
were 40 000 beds in military hospitals where the wounded had the most advanced
treatment such as the celebrated penicillin. As Hippocrates said long ago :
"If you aspire to be a surgeon, join the army and follow it everywhere".

Excepting atomic weaponry (thank God) Normandy saw the use in welfare of every
other technological advance.

In the pages that follow is an account of 50 discoveries chosen at will from many
others, that, in giving the reader a wider understanding of technological advance,
sheds new light too, on the scale of Operation Overlord. Visitors to the Landing
Beaches may thus appreciate how humanity as a whole has profited
from all the inventions collated in the Battle of Normandy.

Some good comes from every misfortune.

ERSATZ -TO REPLACE THE REAL THING

Hitler might have said to Mussolini : "I will teach you how to make butter from coal, when you teach me how to make pullovers from macaroni". This is a witicism, knowing that American chemists were at the time producing a woollen thread from a soya bean. Whilst the main force of industrial and agricultural production was directed to military uses, it must be said that, as with rubber and petrol there was a need for every other product to have a synthetic counterpart.

No limits are set to the genius of man when necessity becomes the rule. German chemists therefore, in the Imhausen factories at Witten in the Ruhr succeeded in producing synthetic fat from coal in the ratio of 70 tons of coal to one ton of nutrient. Production rose to 2000 tons a year. This ersatz butter was so delicious that it was reserved for the most favoured children of the regime : the sub-mariners.

Sugar was produced from wood with a bi-product "furfural" not only used in the production of synthetic mineral-oil and rubber but also for compounds such as Nylon. The sugar could also be used to make yeasts that in turn made possible… A German firm, Milei went on to an egg substitute with a milk and skimmed-milk base. All these ersatz products : sugars, fats, alcohols, yeasts, soaps… were exported to all the occupied countries of Europe in exchange for the real thing. The genuine article went straight into use in the German war effort. The Allies did not remain idle as we have seen regarding mineral-oils and rubber but in foodstuffs progress was more significant. The approach was different : the famous K-ration for GI's were biscuits with a soya base and aimed to replace Corned Beef. In the event of a scarcity of water in Normandy, soap that would lather in sea-water was produced.

If synthetics had gained their reputation during the Second World War the USA then altered course completely to grow vegetables without soil to feed their troops. Tomatoes and lettuces were grown

American "K" Rations from maize and soya supplemented with synthetic vitamins

on gravel in a solution of nutrient and selective weedkiller, under arc-lamps ; this was hydroponic cultivation.

These two advances : synthetics and hydroponics were to be developed considerably later.

HYBRID MAIZE

Maize is a cereal, native to the North American Continent that can be traced back to prehistoric times in Mexico and New Mexico. Christopher Colombus brought back the grain to Spain on returning from his first voyage in 1483. Back in the Thirties the Americans created a hybrid form of maize in a cross fertilisation that was entirely artificial. The result was to bring out the desirable qualities of the strains involved. The disadvantage however is that these qualities are not passed on and fresh seed must be produced for each crop sown. Following their entry into the War, in order to increase their beef, poultry and milk yield… but also to augment the latent potential in maize : plastics and synthetic rubber elements, the USA launched extensively into the culture of hybrid products.

Artificial Pollination

By 1945 up to 100 quintaux (5 tons approx) per hectare, records showed, had been harvested in Indiana and in Iowa, where 100% of agriculture had become hybrid. Production had risen from an average of 14 quintaux in the Thirties to 35 quintaux in 1942 to 1945. Shortly after the Liberation, a misunderstanding of the subtleties of the language of Uncle Sam led the French, expecting wholesome wheat bread to an import that changed the greyish tint of their war-time bread to bright yellow. France had, in fact received shipments of corn flour instead of wheat. Under the Marshall Plan France was to import 30 tonnes of hybrid seed-corn that would produce up to 52 quintaux to the hectare instead of 10 quintaux. The usefulness of hybrid maize had been fully demonstrated. In 1950, 1000

tonnes of seed were imported. Following an agreement between the USA and France, INRA produced French hybrids that are now in current use. The result has been a ten-fold increase in production.

NESCAFÉ-NESTEA

INVENTOR : Nestlé
DATE : 1937

The coffee habit began about the year 850 in Abyssinia. The goat-herd Kaldi, was surprised to find that his goats were particularly enlivened after eating certain red berries. Testing them himself he noticed their stimulating effect. After this, his fellow countrymen, given the effect it produced, began to cultivate coffee. This may be a mere legend but is a story worth telling. Coffee first departed from Yemen to conquer Europe via Constantinople and the Austro-Hungarian Empire.

France promoted its cultivation in her colonies, particularly in the West Indies from which it spread to Brazil.

In the Thirties Brazil had such over-production that coffee was burned in railway engines and Nestlé was asked to take a part of the excess crop and to renew its research for a soluable coffee which, so far had not proved satisfactory.

First attempts failed to retain the aroma of full-bodied coffee. In 1937 by adding carbohydrates, present normally in coffee, Nestlé obtained a soluable powder that was named Nescafé and duly patented it. In 1938 Nestlé similarly produced a tea : Nestea.

Results commercially were disappointing in 1938 and 1939. The first large order came from the American Army and where commercial publicity had failed the GI's succeeded worldwide. It was an undreamt of start. The magic powder arrived on the Normandy beaches where it is still possible to taste Nescafé of that time, without a shelf-life date but with all the aroma of coffee intact. In May 1945 when the American troops met the Russians, the former offered packets of Nescafé to the latter who, in ignorance consumed it undiluted. It is said that the effect on the Russians was similar to that of Kaldi the goat-herd and in some cases soldiers had heart attacks.

SOYA

A Jeep steering-wheel with a plastic distributor head emanating from Soya

Soya is a leguminous plant known to the Chinese for over 4000 years. It was introduced into the USA right at the beginning of the 20th century where its cultivation spread rapidly. First statistics appeared in 1924 with a production of 135 000 tonnes. At their entry into the War, USA production topped 5 million tonnes of Soya beans and its use as food for both humans and animals including other uses was already current. At the same time the exploitation of its pharmaceutical use had already advanced. Richer in calories and proteins than beef, it was claimed that 500 grammes of Soya flour contained as much protein as a kilo of beef. The vitamins A, B1, C, G, K, being present in Soya it can well be understood why the US Army issued their famous "K" Rations in biscuits eaten by people in Normandy at the time of the Liberation. Elsewhere, a number of food substitutes such as oil and fat were Soya based. Soya therefore replaced pea-nut oil in American margarine. The most remarkable achievements however, were in the industrial use of Soya. Soya-based plastics were used in the construction of cars and aircraft. It appears that both the steering-wheel and the distributor-head of the Jeep were cast in a plastic extract of Soya. Yarn similar to wool could be spun into very supple thread. Much else might be mentioned : insecticides, inks, rubbers... In addition, the residue after the extraction of oil, formed a by-product, extremely useful for cattle and poultry food at such low cost that it soon became the main food base for European use on farms.

Synthetic wool from Soya

LAMINATED WOOD -GLUED AND IMPREGNATED

I f one of the first com-
posite developments
was unquestionably
plywood, wood glued and lanimated was
a notable advance. The process involved
the cutting of thin sheets glued together
with the grain of each at different angles
to avoid warping. It was used mainly in aircraft
production. If in the Thirties aircraft construction in metal
was deemed great progress during the War the cost
and scarcity of the metals encouraged the return
to wood and the development of a laminated product.
Impregnated with synthetic resin this provided continuous
sequence as did the newly introduced contact glue,
rendering the material impervious to humidity and organi-
cally resistant. The advance in the use of wood served
for the tail pieces in aircraft and for the entire fuselage
of the twin-engined de Havilland Mosquito that consisted

**De Havilland Mosquito
"The Wooden Wonder"**

of layers laminated -glued, hard-wood and balsa.
This brings to mind the exploit of the Mosquitoes
of Group Captain Pickard of the RAF and Colonel Livry
Level, who on 18th February 1944 brought down the per-
imeter walls of the prison at Amiens to liberate
258 prisoners condemned to death by the
Gestapo ; appropriately the code-name was
Operation Jericho.

Wood de-grained, impregnated and compressed,
becomes another substance that can be used
to make mechanical parts, such as gears,
also electronic and electromechanic components.
Today, not only are there magnificient timber
structures in this material but the interior fittings
of cars and even in car
construction. Its qualities :
light-weight, corrosion-
free and the ease with
which it can be recycled
make it very attractive.

**The wooden
wonder**

Spitfire propellers in impregnated wood

SYNTHETIC RUBBER

from petrochemicals or from butadiene originating from grain alcohol increased. The United States of America formed in 1942 a pool of private enterprise known as G.R.S. : Government Rubber System, under Federal control.The Germans too, deprived of hevea rubber from the Far East, due to the Aliied naval blockade, took similar action.

In Normandy, Allied transport, like that of the Germans, was equipped with synthetic rubber tyres.

The Hevea tree in the Retort

Today every tyre manufacturer uses synthetic rubber which accounts for over 80% of rubber used in tyres.

In 1880, G. Bouchardat prepared a synthetic rubber using isoprene which he had extracted from natural rubber ! What is the use, one may well ask, of this discovery "to make synthetic rubber from real rubber ?" Four years later, rubber would be made from isoprene but extracted this time from turpentine.

At the beginning of the 20th Century a number of establishments invested in projects aimed at perfecting the production of rubber from hydrocarburate.

I.G. Farben in Germany made Buna, invented by Bock with Butadiene (also a hydrocarbon C4H6, of the methane series) in reaction with sodium and obtained a synthetic rubber of a quality superior to the real thing. Charles Goodyear, founder of the Goodyear Company invented the process of vulcanisation in 1839 (the treatment of rubber with sulphur compounds to improve its strength). The Company acquired the British patents in 1927 and produced the first tyres in synthetic rubber, S.B.R. Styrene Butadiene Rubber. In 1930 two Americans J.A. Nieuland and W.H. Carothers discovered Neoprene : an oil-resisting and heat-resisting synthetic rubber made by polymerising chloroprene. After Pearl Harbour the Far East rubber plantations were occupied by the Japanese. America, the most motorised country in the world applied rationing and restricted circulation to limit the use of tyres. Research for rubber

SYNTHETIC MOTOR-FUELS

INVENTORS : F. Fischer and H. Tropsch
DATE : 1923

If today one thinks of great scientists or inventors typical of the creative spirit of the 20th Century, those called to mind might be : Fleming, Turing, Grignard but few would mention Fischer and Tropsch.

And yet Franz Fischer and Hans Tropsch doing research at the Kaiser Wilhelm Institute for Kohlenforschung in 1923 discovered how to produce synthetic hydrocarbon from coal gas that would revolutionise the production of motor-fuels into the 21st Century. For the first time ever Fischer and Tropsch effected the means of turning gas into liquid. To obtain hydrocarbon H2 is brought into contact with CO. CO will have formed when either carbon or natural gas has been brought into contact with oxygen.

The compound H2-CO is passed through a catalyst to become liquid hydrocarbon. That catalysts could produce a chemical change without the elements being consumed had been known at BASF since 1910.

The first industrial plant to produce synthetic motor oil was set up in Germany in 1935. During the War

From one fossil to another

9 factories were opened, capable of producing 16 000 barrels a day and estimates are that in 1944 production in Germany would have risen to 4.5 million barrels of synthetic motor-fuel. Cobalt was first used as a catalyst but as it became scarce, iron was used instead. As is mentioned elsewhere it was the perfecting of their logistics that assured the Allied victory, but also by destroying the logistics of their adversaries. This was patently so regarding enemy oil production. After the War two synthetic oil plants were dismantled and set up in the USA as models to test their performance.

Similarly, in South Africa during the sanctions against apartheid two plants were set up using the large deposits of coal from open-cast mines. Today with the price of a barrel of oil at $20, oil from coal has enormous possibilities. The need to drill for oil at great depths increases production costs. Reserves are limited, whilst resources of gas and coal are enormous. On this basis the production of synthetic oil can well be envisaged near coal fields and sources of natural gas.

NYLON

INVENTOR : W.H. Carothers
DATE : 1930

H. Standinger

In 1922 the German chemist and Nobel prize-winner, Hermann Standniger (1959), discovered the principle of polymerization, namely a large molecule formed from repeated units of smaller molecules and known as a macromolecule. In 1928 the Du Pont de Nemours Company founded in USA in 1802 by an imigrant French chemist to make gunpowder, decided to diversify and engaged a research team among whom was a brilliant Harvard chemist : Wallace H. Carothers. In 1930, when allowing mixed molecules to heat up for longer than intended, W. H. Carothers discovered by chance, a new polyamide from which threads could be made and was to become after long research : Nylon patented in 1937.

Due to its fine and light texture it was immediately used for ladies stockings. From their entry into the War the U.S. Army, knowing its qualities requisitioned the total production for military use. From the first, nylon being stronger and lighter than silk was used to make parachutes.

From stockings to parachutes

These were first used in the Pacific at Midway in 1942. On 6th June 1944 at dawn, thousands of Allied paratroops, using nylon parachutes, descended from the skies of Normandy to secure the flanks of the landing bridge-heads : the Americans at the base of the Cotentin and the British in the Bay of the Orne.

A few days only, sufficed for girls in Normandy to make blouses and skirts of this new light and stylish material in many colours. Parachutes had been dyed in colours denoting their various uses. It is rather difficult today to imagine life without nylon.

CHEMICAL ENGINEERING

REINFORCED PLASTICS

INVENTOR : Du Bonnel

If plastics in general are known as an important development between the two wars, notably in aviation, their use with fibres has also produced new materials known as composites. Composite materials are produced by a process used in nature itself. They are formed by a matrix but the fibres that go to the making of the composite bring to it qualities not possessed by the matrix. An example of this is that of wood fibres immersed in lignin.

From the Iron-Age to the epoch of fibres with intelligence

Fibre materials were developed right from the beginning of the War. These were mainly forms of fibre-glas invented by a Frenchman named : Du Bonnel. Glass fibres coated with a matrix of polyesters, also wood fibres were impregnated with synthetic resin.

The qualities of fibre-glass : its light weight, resistance to humidity, its strength allied with performance

in varying temperatures, make it ideal for production of reinforced plastics using either long or short fibres. In USA, Britain and in Germany, reinforced plastics were employed in the making of components that were light-weight, had greater shock resistance and toughness than metal. In addition reinforced plastics were cheaper and quicker to produce. In view of these qualities, reinforced plastics were first used in aeronautics and electronics : aircraft tail-panels, fuel-tanks, propellers and aerials...

In England recently, apart from the usual fibre glass, long optical fibres have been used in composite; throught which, is passed the ray of a laser, making it possible to detect constraints. The first thinking composites had been born !

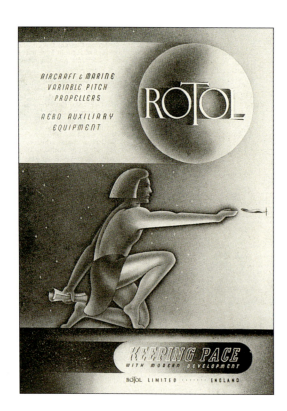

Rotol : The first British company to produce propellers in reinforced plastic

SILICONES

V. Grignard

At the end of the 19th Century a number of savants were doing research on Silicium, one of them : Kipping, in Britain, hit on what was to become "Silicones". The mathematician Victor Grignard born in Cherbourg in 1871, was working on Sicilium in Lyons and discovered the famous test that was named after him. It was a radical advance and brought him the Nobel Prize in 1912. As professor in Besançon, Grignard might have forgotten his native Normandy. The Great War however brought him back to Cherbourg when he was called up and found himself in the highly strategic post of guarding a level-crossing ! Grignard's discoveries led Kipping to devise a whole family of silicones in the 1920's and the uses to which they could be put.
Their practical use did not come until the beginning of the Second World War when two great American industrial groups : Dow Corning and General Electric realised their importance.

The prodigious properties of insulation and as a moisture repellant were basic to their most important military uses : the protection of sparking-plugs in aircraft at high altitudes, the encapsulation of electronic equipment like Radar, for use as a demoulding agent when turning out casts both in the rubber industry and for foodstuffs. Other uses where information as to their use was under wraps, included silicone oils in aircraft shock-absorbers adding a dimension to their performance produced by the action of molecules in the substance sliding against each other. The War witnessed considerable development in the uses of silicones notably due to the German scientist Müller and the American Rochow.

> *Chemical engineering revolutionised by Cherbourg Nobel prize winner*

The annual production of silicones now exceeds a million tonnes with such diverse uses that it is impossible to list them. The bio-compatibility of silicones renders them especially useful in the making of implants and surgical accessories, kitchen ustensils and for silicone-based anti-adhesives, rendering the material moisture resistant.

ENIGMA & THE COMPUTERS

INVENTOR : Hugo Koch
DATE : 1919

Crypto-analysis or how to decypher enemy secret communications has always existed. The Spartans in 40 B.C. used the "Scytale": a baton with a thin strip of parchment which, carried a message. When taken off the baton the contents of the scroll were indecypherable by the messenger taking them to the recipient. The latter however, had an identical baton which, once the parchment was placed on it revealed the secret.

Hitler in 1934 had a coding machine made that coded on an entirely new principle and was considered to be indecypherable : this was the coder Enigma, the 20th Century Scytale. It was really a copy of a commercial machine invented by a Dutchman : Hugo Koch who patented it in 1919, which he later ceded to a German who called it Enigma. It had no success commercially but taken over by the military it was the supposedly unbreakable code. The Polis Secret Service, having acquired a commercial version of Enigma evolved with the work of eminent Mathematicians an analytical method of decoding. Concurrently in 1937 their French counterparts under Colonel Gustave Bertrand brought pressure to bear on a German who "revealed" the secret of Enigma, for cash. France could, from then on, decode all the German messages.

Finally a Polish Jew known by the name of Lewinski, who had worked with the Germans on the Enigma Code machine informed MI 6 in June 1938 that he had obtained all the secrets of Enigma from a repliqua he had made himself. MI 6 sent an expert in code and a young mathematician : Alan Turing to Warsaw to take note of what they had learnt from Lewinski. This being done, the British Government decided in 1938 to construct an automatic decoding machine following Turing's conclusions; this came to be known as "The Bomb". It was

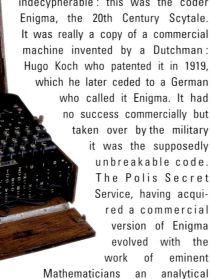

Enigma : millions of combinations to encode a three-letter word

The inconvenient genius

hardly what could be called a computer but there was strange affinity. So on 9th January 1939, the Poles, the French and the British, all of whom knew the Enigma Code decided at a meeting held at Vignoble in France to coordinate their efforts and send their cypher specialists with their equipment to Bletchley Park (north of London) to avoid their falling into enemy hands. Up to then, France had already decoded 141 messages allowing the British thanks to " The Bomb " to decode 15 000 and strange as this may seem Poland, France and Britain were taken unawares. This without all doubt is the greatest enigma of the War. On 14th November 1940 a message on Enigma, decoded, revealed the impending bombardment of Coventry. Churchill being informed, had to decide whether to evacuate the city, revealing to the Germans that the code had been broken, or allow them to carry out the raid; he chose the latter. The number of messages to decipher growing

to around 10 000 a day, Flowers and Chandler were given the task of devising a more powerful decoder. It was the Colossus mentioned in an article by Turing on 26th May 1936 in which he described an unprecedented machine,automated, rational, abstract, all-embracing, capable of working in conjunction with other machines, having instructions and information that would appear on tape with the answers after which the machine would stop on its own. At the same time Turing, a pupil of Einstein at Princeton worked with von Neumann and whilst it will never be known what influence one had on the other it is unquestionable that Turing is the father of the computer and artificial intelligence. During the War, Turing and Neumann met in secret. Chandler and Flowers believed that, at the same time as Colossus, Turing produced an electronic and electromechanical computer. In the night of 25th to 26th June 1944 the Germans sent a message coded by Enigma that an armoured column was advancing to Pegasus Bridge. They were unaware that this was an Allied sound-

Alan Turing (1912-1954)

illusion. On the evening of 2nd August an Enigma message from Juvincourt described E. Sommer's reconnaissance in his jet plane over Normandy, indicating all the Allied positions. One could continue indefinitely with examples of German coded messages, decyphered immediately by the Allies.

When the War ended, to keep secret the reason for Allied successes, Churchill ordered the destruction of Colossus and similar apparatus. He thus destroyed a great deal of technical aptitude which might have been handed on for industrial use. In restitution for this outrage in the History of Science, Colossus was reconstructed in 1999; unfortunately this did not bring Turing back to life, for if their machines were destroyed most of those who used them "disappeared" mysteriously. Turing who, was alleged to be unbalanced, was condemned for homosexuality and "committed suicide" on 7th June 1954.

The genius of Turing is now internationally acknowledged.

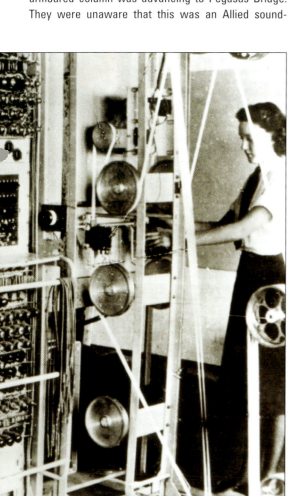

The computer: Colossus. Note the paper tapes with transmission by perforation

PERMANENT MAGNETS

INVENTOR : Mishina
DATE : 1931

The Lodestone, a natural form of the magnetite has been known since Antiquity as an artificial magnet. In 1600 William Gilbert tried to make artificial magnets using field polar magnetism. The discovery by Oersted in 1820 of the magnetic field created by an electric current put a brake on attempts to produce artificial magnets. Not until 1931 did the Japanese Mishina discover an alloy of iron, nickel and aluminium, that had the exciting properties of a permanent magnet. It was named : "Alnico".

In 1938, Olivier and Shedden, in the course of the cooling process applied a magnetic field to the alloy and obtained some remarkable performances.

Alnico was found to have excellent magnetic properties and great durability that made it an essential component for all military systems developed during the Second World War by all the beligerants.

Artificial Magnets can now be found in every branch of industry : electronic/electro-acoustic, radar, loud-speakers, microphones, gyroscopes, telephones, teleprinters, tape-recorders…

- electrotechnology, generators, magnetos, and magnets motors, relays, switches, contact-makers, butterfly-valves…

- instruments of measure :

- ammeters, voltimeters, electric comptometers or speedometers, clocks…

Thanks to the work of Louis Néel from 1944-1946 it became possible to make artificial magnets using metallic powders which proved a simpler method than hitherto. L. Néel was awarded the Nobel Prize in 1970.

An advertisement by Philips for Ticonal magnets

LONDON NEWS   JUNE 24, 1944

Magnets

He earns no medals ; his name will never make headlines ; his way of life is modest and his work unexciting.

But to us he is an individual, a personality ; not just a number on the time clock. He is a skilled man — and more, because for years his skill has been allied to the Philips tradition of doing things more efficiently ; of making things just that much better.

He melts metals — very special metals which are used for making ' Ticonal ' * permanent magnets of unusual power and unique properties ; an outstanding Philips invention. He is one of the thousands of Philips workpeople who gave you, before the war, the Philips products you knew and trusted so well. His skill is a vital asset to the nation today.
* Registered Trade Mark.

PHILIPS
RADIO ★ LAMPS
AND ALLIED ELECTRICAL PRODUCTS

PHILIPS LAMPS LTD., CENTURY HOUSE, SHAFTESBURY AVENUE, W.C.2.   (24J)

THE HANDIE-TALKIE

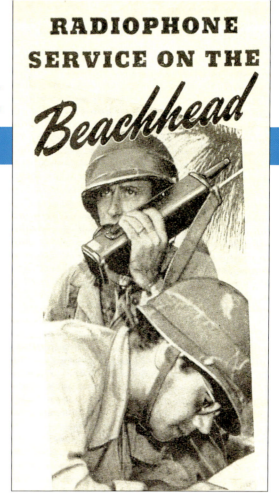

The Dane V. Poulsen and an American R. A. Fessenden invented the forerunner of the radiotelephone in 1902. In 1940 Motorola examined the matter of producing a portable radio telephone, a two way transmitter-receiver, that is, one could speak and listen simultaneously with a given frequency modulation : Handie Talkie. The first completed system was put in place for the Bowling Green Police, Kentucky in 1941. To be carried in one hand, the Army Handbook stated that it should not weigh more than 5 lbs and have a range of between 1.6 and 4.8 kilometres. The Handie Talkie worked on a band of frequencies between 3.5 and 6 MHz. One of the frequencies mentioned was crystal controlled. Within a very small space little larger than a telephone, $30 \times 8.5 \times 8 \text{ cm}^2$ it contained 5 vacuum tubes; transistors were not invented before 1948. It was a perfect pearl of miniaturisation. The tubes required an HT battery of 103,5 volts and the filaments needed a 1.5v battery. The batteries alone weighed 900 grammes, that is, almost eight times that of a portable telephone today : 110 grammes.

Always smaller always further

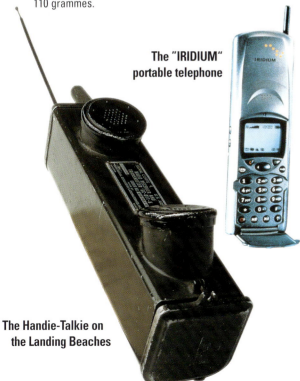

The "IRIDIUM" portable telephone

The Handie-Talkie on the Landing Beaches

All the American Units landing on D Day were equipped with the Handie Talkie. Even today it is not unusual to find one in working order, if one is lucky enough to have the right batteries. This forerunner of the portable telephone disappeared with the arrival of transistors. The expertise acquired by Motorola provided a forward leap. A notable advance by this Company was the Transponder that provided Niels Armstrong with a relay from Moon to Earth in 1969 " A small step…". It was Motorola too that won the race with "Always further, always smaller" in producing the first portable satelite telephone that could communicate with all parts of the world : the Iridium system. A constellation of 66 satelites in low orbit, put into Space by the rockets Delta, Long March or Proton, joined up a network of 11 stations on Earth making telephone communication possible from whatever position on land, sea or in Space. It may be noted in passing that 66 plus 11 makes 77, corresponding exactly to the number of electrons gravitating round the nucleus of Iridium.

Motorola would not remain for long uncontested ; From the beginning of this century, Global Star (Loral-France Telecom), Skybridge (Alcatel) and Teledesic (Microsoft) will be ambitious competitors.

MILITARY DECEPTION

In January 1941, to demoralise the Italian garrison besieged in Bardia in Libya, the British Army, using a gramophone and loudspeakers simulated tank movements in the night, that induced the Italians to surrender at first light. This concept of psychological warfare was developed within the framework of Overlord, notably by the Royal Engineers who specialised in producing items of military deception. They were based at Colombiers-sur-Seulles in Calvados. The "Deception Unit" made fake tanks and fake planes on faked airstrips… to deceive the Enemy. Whatever remains of the Unit's time there, can still be seen at Colombiers in the shape of Eros of Tierceville, the work of an architect and Paris theatre scenic-artist Major Papillon who joined the British Army and accordingly left his mark.

The Libyan experiment had its limitations so the use of hi-fi cinema projectors were envisaged, developed by Metro Goldwyn Mayer in 1929, for open-air use in Hollywood. It should be remembered that in 1944, tape-recorders only existed in Germany and magnetic sound was not used

The Art of Pretence

Doll with an explosive charge dropped by parachute in the Cotentin

in USA cinemas. During the night of 25th to 26th June the Unit contrived to simulate the sound of a column of heavy armour by mounting on a scout car two cine-projectors with film sound-tracks running at full blast, giving the impression of a tank column moving from Perriers-sur-Le-Dan and over Pegasus Bridge. A decyphered Enigma message proved the strategem to be a complete success. It was repeated therefore with similar success later at Fontenay-le-Penel, Rauray and Gouvix. The quality of sound reproduction was hardly that of Imax stereo but rather a kind of dimension to sound.

Inflatable Sherman Carried by troops

THE TAPE-RECORDER

I n 1897, a Frenchman Janet, described a possible method of sound reproduction by magnetising steel wire. In 1898, a Dane Poulsen described and produced a "Telegraphon", consisting of steel wire rolled round a cyclinder, similar to the drums used in Edison's phonographs. The wire recorded magnetic signals from sounds through a microphone and played them back. Poulsen's invention was exhibited at the Universal Exhibition in Paris in 1900. Over the years, tape-recorders on the steel-wire principle were updated. It was quite usual to find them in crashed aircraft of all nationalities in Normandy in 1944. On the German side, Pfleuneur patented his invention, consisting of a paper band with a thin coating of iron filings that retained magnetism.

In 1933, BASF expressed an interest in the latter and joined with AEG to produce magnetic heads consisting of an electro-magnet of which the air-gap was a fine slit causing the magnetic field to open out.

In 1935, BASF subtituted the powdered steel for a much finer magnetic film, produced by a chemical reaction. AEG, BASF and Telefunken combined to produce

Tape-recorder AEG, 1936

the "Magnetophon" as we now know it. The first concert recording was by the London Philarmonic Orchestra under the direction of Sir Thomas Beecham given in BASF's own hall at Ludwigshafen on 19th November 1936. During the War, Germany mobilised a considerable amount of research at AEG Telefunken in the development of the tape-recorder which provedto be a remarkable propaganda weapon in the hands of the Nazi regime.

From sound-track to magnetic-tape

In 1941 the first professional tape recorder was exhibited at "the Universum Berlin Film AG". The few copies on sale were rapidly sold. The Allies noticed that Hitler's speeches and concerts were often broadcast by local radio stations which led them to believe that recordings were distributed throughout the country. Astonishment at the high quality of these transmissions was world-wide. Among those astounded was a certain J. Mullin, on location in London as representative of a Californian cinema organisation. After the War he managed to get hold of some tape-recorders seized by the Allies and passed them on to Ampex who made copies. The era of tape-recorders, cassette-players, videoscopes and the like was born.

AERIAL PHOTOGRAPHY

INVENTOR : Nadar
DATE : 1858

It is to Nadar, alias Felix Tourmachon that we owe the first aerial photo, taken aboard a balloon at 520 metres altitude above the Avenue du Bois de Boulogne, in Paris in 1858. The negative, still extant is kept at the CNAM and the Place de l'Etoile is clearly visible on the right. In the same year Nadar took out a patent for : "A new aerostatic photographic device ". If Wilbur Wright is credited with taking the first photo from an aircraft the exact date or place is not known. It may be either at Auvours near Le Mans in 1908 or at Contocello in Italy in 1909.

The Great War certainly lent wings to the development of aerial photo reconnaissance. It became possible to locate troop movements in the Battle of the Marne. Cameras and photographic material were constantly improved. Pictures once taken manually could later be taken by remote control. Photos could be taken, not only on planes and airships, but from balloons and even from kites.

Between the Wars, the Allies abandoned their efforts at improving reconnaissance aircraft to concentrate on better maps and map-reading or geology. A few architects like Le Corbusier used aerial photography to plan urban lay-out. In the same period the Germans used commercial aircraft to photograph their future prey : USSR but also France and England. The French and British, taken unawares, put into battle ill-adapted reconnaissance aircraft, suffering heavy losses as a result. An Australian, S. Cotton, saved the situation by developing a new strategy. He used unarmed Spitfires, being lighter, speedier and at high altitude : 10 000 metres. The only means of defence available for these pilots was their consummate art of manœuvre. S. Cotton had no losses.

Whilst it was vital to locate Enemy troop positions and defences in spite of camouflage and deceptive tactics, it was also necessary to discover military and industrial potential. Amongst the more outstanding photos may be noted one by F/Sgt. E. Peek. From his Reconnaissance Mosquito, he took pictures on 23rd June 1943 of a V2 in take-off position on the firing-line at Peenemunde. In the preparations for Overlord, millions of negatives were used, not only for map-making but to identify obstacles on the beaches, defences along the Atlantic Wall... around 26 000 negatives daily, needing 60 000 prints. The demand was so great that the British photographic industry soon ran short of the silver

The first aerial photo in the world : taken aboard a jet-plane Ar

DT/SP 1/France, N.Asnelles-sur-Mer/63A
49N 1W

over Arromanches on 2nd August 1944

compound needed to make the emulsion. Supplies were sent by air from the USA. Use was made of both civil and military geographers and geologists to examine the photos and advise those involved in Military Intelligence. Amongst these was a certain Ian Fleming, later to be "the inventor of James Bond". Although it can be said that Lower Normandy was the most photographed place in the World, a certain amount of confusion existed, in spite of all the care taken; the hedges in the Bocage were an unpleasant surprise to the Allies.

In the hectic scramble to go faster, to attain greater heights, it was in the event, the Germans who gained world admiration: with the first jet aircraft Arado 234, bi-reactor.

See first: then attack

The plane's mission was to photograph targets for the V1 and V2 rockets and to correct the shots after the first impacts. The plane flew at 12 000 metres and attained a speed of around 900 km/h. On 2nd August 1944, behind schedule due to Allied raids, though also due to action by the Resistance the first Arado took off from Juvincourt in the Aisne. Aboard was Captain E. Sommer. For an hour and a half he flew over Normandy, unhindered three or four times, with a Zeiss Rb 30 x 50, and a focus of 50 302 millimetres, a camera of high performance. He took 380 remarkable photographs of the entire Allied logistics. The intelligence received and transmitted by Enigma from Juvincourt was to unleash the counter offensive from Mortain that same day, shortly before midnight. .
Following the perfection of the Arado Jets came not only the U2's but also the Drones, pilotless planes (self-propelled aircraft) that remained airborne, anything between a few minutes and several hours. There was also a kite with four cords that could carry cameras even radar. The first negatives in Space were taken in 1946 at an altitude of 130 km aboard a V2 rocket fired in USA.
Today satellites transmit numbered pictures of the Earth for many uses : environmental, statistical for areas planted or forested, flooded zones, geological, archeological research...
The first satellite photo of the Earth dates from 1959.

(See : Normandie 44. Les photos de l'avion espion.
P. Bauduin et B. Charon, Caen 1997)

INTERCOMMUNICATION

PHOTOS AND FILMS IN COLOUR

INVENTOR : G. Lippmann (N.P.)
DATE : 1891

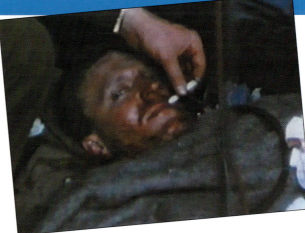

The first photograph in colour was obtained by a Frenchman: Gabriel Lippmann in 1891 using interferential waves of light. He was awarded the Nobel Prize in 1908 and elected President of the Academy of Science in 1912. Not until the First World War was there a breakthrough in films. The Americans: Leopold Godowsky and Leopold Mannes perfected a method using bichrome for use in cine-films. In 1930 with the Kodak Company they succeeded in producing a film for use by the amateur photographer. Five years later they brought out a trichrome film commercialised as Kodachrome. In 1928 another American: H.T. Kalmus invented a new form: Technicolor. In 1937 Agfa in Germany put Agfacolor on the market a few months ahead of Kodak's Ektachrome. With the competitors lined up, the race had started. In 1939 a new Afgacolor : Negative-positive was perfected. In America, that same year MGM produced its famous record running-time film "Gone with the wind", directed by Victor Fleming. In 1943, after several abortive attempts, Germany produced in Agfacolor "The Adventures of Baron Munchausen" a sort of propaganda film, highlighting technical advance for the 20th anniversary of the Universum Berlin Film Ag. It was produced by J. von Baky. A few magazines at the time had begun to publish photos in colour: "Illustrated" in the UK and "Signal" in Germany. War correspondents did not yet use colour. Colour films at that time,

Extracts of the original film taken on board the B17 Memphis Belle on its last raid on Wilhelmshaven in 1944

required in fact three cameras combined: one
for each main colour. This did not make for easy
transport. Only two colour films in Kodachrome
were produced for amateurs, with 16 mn
cinema-cameras. "Memphis Belle" merits a special
mention. It was the true story of a B17 based
in England, filmed from inside the Bomber in a raid
on Wilhemshaven. It depicts thrilling scenes
of aerial dog-fights. The series of films by George
Stevens produced by per-
mission of Eisenhower are
the only colour films on the

Agfa versus Kodak

Liberation of France.
Of note are striking pictures
of the Landing Beaches
at Bernières, in Caen,
St Lô, Coutances, Carpiquet
and Paris... These films, retrieved after the death
of the producer, were published by his son, only
a few years ago.
Afga's rights were dispersed after the War and
have been used to make up current present-day
usage.

G. Lippmann

RADAR

Home chain Radar on the East Coast of Englan in 1940

stand out: Sir Watson-Watt of the N.P.L., Dr R. Kühnold and Maurice Ponte so that in 1934 CSF applied for a patent for "un nouveau système de repérage d'obstacles et ses applications" a new system of identifying obstructions and its functions. This device was used notably on a liner, the "Normandie" in 1936; not exactly radar as yet, but all the components were there.

By the end of the Thirties, the British, the Germans and the French too, possessed a range of devices: extremely efficient ones, not merely to locate but to fire - on targets.

Right from the beginning of the 2nd World War the British erected large radar aerials: the Home Chain to protect their coasts against attack from German aircraft. It is now a fact that RAF pilots and Radar won the Battle of Britain.

From the Titanic to micro-wave oven

On 8th May 1940, Maurice Ponte, an engineer with the SFR now THOMSON-CSF brought to the British, his invention: a transmitter-valve that proved a break-through, the magnetron that was to revolutionise the development of Radar. In June 1940 on the Côte d'Azur an SFR French Radar detected a flight of Italian bombers, largely destroyed therefore by French fighters.

Radar could henceforth become mobile, notably aboard aircraft. Philippe Livry Level, a Flying Officer from Normandy who joined the RAF tells how, thanks to Radar he could pursue enemy submarines in the Atlantic in 1942.

This brings us to Normandy where the German occupants installed two Radar surveillance networks: one naval and the other for aircraft. Some thirty aerials were set up along the Normandy coast-line for both detection and firing-points. They were of four different types on the following sites: Dieppe, Fécamp, Cap-d'Antifer, Bruneval, Pourville, les Andelys, Theil-Nolent, Houlgate, Douvres, Arromanches, Ducy-Sainte-Marguerite, Englesqueville, Saint-Pierre-Église, Clitourps, Cap-Levy, Urville, Auderville-Hague, Jobourg, Beaumont-Hague, Sortosville, Saint-Pierre-la-Vieille, Vire...

Six hours after the D-Day Landing the first mobile Radar was in use in the British Sector.

The British landed some 300 Radar-equipped trucks in Normandy and since the Americans brought in

The history of Radar is a complicated one. Since Hertz's experiments in 1886 his findings are known in every country. Heinrich Hertz discovered that a radio-electric or optic wave reverberates against an obstacle and that the echo emitted can be recorded.

The sinking of the Titanic first incited research for devices to locate obstacles or objects in the sea.

An Englishman in 1912 invented a device to detect echoes in water: the Sonar or Echo-sounder; it was the precurser of Radar.

During the Thirties thousands did research the world over, on angles or reflexion of sound waves on ships and aircraft. They worked on the basis of what later became Radar. It is impossible to say who actually invented what is a collective work, but three names

Radar SFR On the liner Normandie in 1936

at least that number the Normandy countryside can be said to have bristled with aerials.

Radar was in use just as much for navigation and weather forecasting.

In the Thirties radio-navigation depended on beacons on land or radio beacons like the one at Ver-sur-Mer that could give an aircraft its exact position. This was an elementary system of direction-finding. To the three Allied systems used: Loran (USA) in hectometric waves, the British GEE in metric waves and DECCA (also British) but in kilometric waves and the German system, X Gerat, Radar was soon to be added.

During the landings, the Allies used Radar to guide in their gliders and the Resistance themselves used Radar to direct parachute landings. This was the "Eureka-Rebecca" system mentioned by Livry-Level of 161 RAF Squadron in his secret missions. The landing-marker called Eureka was either dropped by parachute or put in place by men on the ground. It sent out a signal in reply to that received by Radar Rebecca from the aircraft. The system was used by 300 planes on D-Day.

There was rivalry the world over, in ingenuity to develope various sytems of Radar both on the ground or aboard ships and aircraft, all in use during the Battle of Normandy. The estimated cost of research and development of the

different types of Radar was 3 thousand million Dollars, costing the equivalent therefore of the German V1 and V2 rockets. After the War many firms continued to develope Radar. Amongst these, Raytheon who had an engineer, Percy Spencer, studying the new magnetrons, in 1946, noticed that a snack placed near one of them had warmed up. The micro-wave oven was born. The magnetron would bring good times to Normandy after the bad times. The wheel of fortune would turn.

German radar at Arromanches in 1944

INDUSTRIAL RADIOGRAPHY

INVENTOR : Wilhelm Conrad von Röntgen (N.P.)
DATE : 1895

W. C. Röntgen

The German Wilhelm Conrad von Röntgen discovered X-Ray in 1895 in Würzburg. He was awarded the first Nobel Prize in Physics in 1901. X-Ray was the only means of inspecting the human body for a long time, first of all the skeleton and later for the organs. Medical generators use up hundreds of kilovolts.

The low voltage limited for a long time, the use of radiography in industry. Not until the beginning of the War did G.E.C. manage to build a generator of a million volts, making it possible to have pictures of large metal parts on their completion. The device was adapted and rendered mobile at the instigation of the USAAF to inspect damage to aircraft on their return from bombing raids. Inspection was carried out on damaged planes by X-Ray. Speedy identification of the damage meant quicker and more precise repairs. The era of speedy verification in the sphere of constructive equipment in industry had also been born.

Inspection by X-ray of an aircraft wing

Inspection of non-destructive components by X-ray

"V MAIL"

The "Victory Post" was, with blood transfusion, an indication of Allied organisation and logistics. Nothing was of greater importance to a soldier than prompt receipt of private mail: the more so, since it implied that Supreme Command's communications were finally in place. The psychological factor alone, merited the efforts needed for an efficient Forces Postal Service. The first Allied planes to land in France, not directly concerned with military action, were Hurricanes of RAF 46th Group specially assigned to transport mail. These aircraft, after the Battle of Britain were stripped of weaponry for lightness and speed; they landed at Saint-Croix-sur-Mer on Saturday, 10th June at 6.30am. Reported thus, the event appears commonplace; it was in fact an exploit. The Landings had begun only five days earlier, during which time the Engineers had constructed an air-base so that those receiving their mail would know immediately that the military situation was stable. Speaking of Sainte-Croix, some noteworthy arrivals took place at this temporary airstrip under under the command of J. E. Johnson. On 13th June two FAFL Squadrons of Spitfires landed: the Free French Airforce; among them Denys Boudard grounded on French soil after a long exile. On 15th June Eisenhower's B17 touched down on this 1 200 metre runway. Last but not least, came the Fiesseler-Storch on 21st June, aboard which was... Churchill.

If the Hurricanes delivered British mail, they also brought messages for G.I.'s known as "V Mail" The messages from their families to soldiers and vice versa were written on special forms. They were censored, coded, sorted and microfilmed; then delivered to the troops. One case of microfilms replaced 37 sacks of mail.

V-mail forms

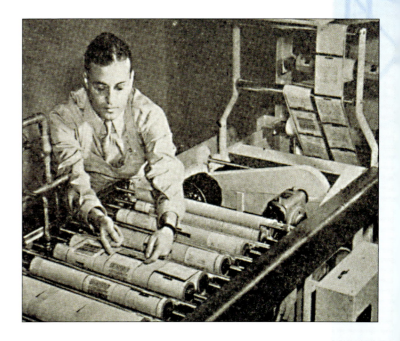

Processing mail

Cable-layer Pluto aground on Urville-Nacqueville Beach

PLUTO

T he meticulous manner in which each sphere of the Landings was handled: military logistics, construction of ports, bridges, airfields and roads... the placing of pipelines for oil or water, the routine of vehicles... rivalled in quality all the Arms of Service on sea, air and land.

Eisenhower, with pen to paper, averred that with the prefabricated ports the "Pipeline under the Ocean" PLUTO was one of the two major innovations of Overlord. During a demonstration in 1942 of a tank flame-thrower, fitted with a long flexible tube for fuel supply, Lord Louis Mountbatten asked the Minister for Fuel, Geoffrey Lloyd whether he would be able to lay a flexible pipeline across the Channel between England and France. So it was that, having already used flexible pipeline to set the sea on fire at the possible attempt at invasion by the enemy that British firms put in place: PLUTO. Amongst these firms was the British branch of the German Siemens, directed by Dr H.R. Wright. The technique finally employed for the pipeline was that of the transatlantic telephone cable. The idea

The Norman Emirate

was simply to reproduce similar cable without its core of copper, leaving an empty tube for oil supply. The scheme proved less simple in practice; supple tubes were used but they were of steel. Two lines were laid: one between the Isle of Wight and Cherbourg with two cables, two pipes with a total length of 280 miles. The other line between Dungeness and Boulogne consisted of 11 cables with 6 pipes with a total length of 500 miles. Pipeline links with Cherbourg and

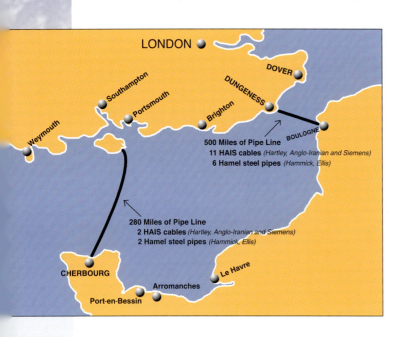

LONDON

DOVER

Southampton

DUNGENESS

Portsmouth

Brighton

Weymouth

BOULOGNE

500 Miles of Pipe Line
11 HAIS cables *(Hartley, Anglo-Iranian and Siemens)*
6 Hamel steel pipes *(Hammick, Ellis)*

280 Miles of Pipe Line
2 HAIS cables *(Hartley, Anglo-Iranian and Siemens)*
2 Hamel steel pipes *(Hammick, Ellis)*

CHERBOURG

Le Havre

Arromanches

Port-en-Bessin

Connecting 6-inch pipelines

pipeline was in action on 12th June at Port-en-Bessin: a provisional terminal in advance of PLUTO. By 24th June the first pipelines came ashore on the sandy beach of Urville-Nacqueville west of Cherbourg and on 3rd July, PLUTO: the flow of oil from England to Normandy began in earnest. Non-stop round the clock, the system pumped 4 million litres of oil daily and from 12th August 1944 to 8th May 1945, 480 millions. If people in Normandy remember pools of petrol in ditches alongside pipes hardly below ground level, leakage amounted to only 1.8%. Astonishing technical success for the time: pliable pipelines and off-shore platforms now in current use in France were then unknown. France is now in the fore front and leads the world in fact, pumping oil from a depth of over 1709 metres, another record.

Boulogne were made possible as the troops advanced. Meanwhile a first cable was brought ashore; it was then connected to the oil in bulk terminal at Port-en-Bessin. Apart from the pipes and cables the vessels needed were part cable-layer, part tanker. The cable-drums, known as "Conundrums" 7 m in diameter, 30 metres long, and 1 600 tonnes in weight fully charged were towed across the Channel by two tugs. In addition pumping stations were set up. The first

Floating cable-drums (Conundrums) loaded with pipeline laid across the Channel

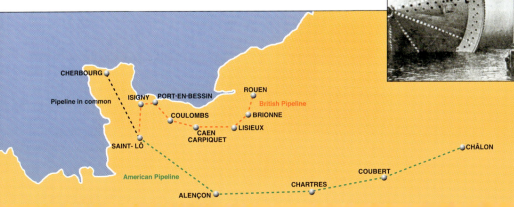

DIVING EQUIPMENT

INVENTOR : H. Fleuss
DATE : 1870

We owe to a certain John Dean of Whitstable in England the first breathing apparatus for use in adverse conditions. In 1830, when his stable was in flames, John took the helmet from a suit of armour in his manor house and attached a tube through which his brother Charles pumped air from a winnowing machine. He thus managed to save his horses. In 1832, he patented the device.

In 1834 he adapted a helmet for diving in the Thames Estuary.

In 1838, Augustus Seibe, an Australian invented the first metal diving helmet fitted with a pump. He supplied the equipment to the French Navy.

This type of equipment, Marguile, is totally dependent on the air supply from the surface and in consequence less mobile. In 1870, Henry Fleuss in England of the Seibe Gorman Company introduced a self-contained breathing apparatus with oxygen. Sir Robert Davis produced equipment to rescue submarine crews in danger, which is continually updated. If the emission of gasses presents no problems for work on submarines generally this is not so for divers and swimmers on active service where utmost discretion is the rule. The emission of bubbles or noise gives away their position. Consequently, in 1942, the Seibe Gorman technicians brought out the first self-contained breathing apparatus that recycled the exhaled gases, hence no bubbles and no sound : "Davis escape apparatus". The toxic CO_2 was emitted into a compartment containing lime sodium. Experiments with this diving gear cost the lives of many divers, due to physiological problems. It could only be used to a depth of 7 metres. The divers and frogmen, taking samples of sand and soil used it on the coasts of Normandy prior to Operation Overlord. On

No bubbles, no noise

New Year's Eve 1943, Logan Scott-Bowden, R.E. landed a boat on the beach at Ver-sur-Mer to take samples of sand, so as to dispel doubts on the feasibility of landing British tanks there. There was no moon on the night chosen but it was also thought that the guards would be distracted by the Festive Season. Unfortunately the lighthouse, not normally functioning, beamed out in celebration.

Our frogman relates how he was forced to co-ordinate his advance with the rythm of its rays. A few months later Davis equipment was used in maintenance on the prefabricated ports and when refloating sunken ships. Lionel Crab used the Davis apparatus.

The Germans too, the Italians and later the French were involved in perfecting the gear now used by divers the world over, both amateur and professional. Advances made in wartime made for further development as much for industrial work in the sea as for scientific. Noteworthy are the number of accessories : diving-suits, pressure-reducers, depth-finders, sound-equipment ... and of course, calculators making longer and deeper diving possible and safer. It is in the composition of the gas inhaled; a mixture of helium, hydrogen and oxygen that the most significant work has been done.

Divers on Active Service with Davis respiratory and exhalatory recycling equipment

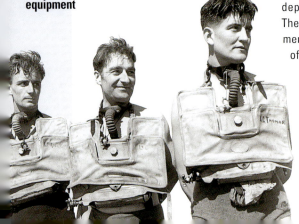

AIR-SEA RESCUE

Life-saving precautions have always been a pre-occupation aboard ships, for passengers and crew. A more recent development has been that of air-sea rescue, notably the specific needs for retrieving pilots forced to bail out. A pilot is in fact more precious than a plane and the treasures of ingenuity included devices to keep them alive.

Almost all aircraft are equipped with inflatable dinghies with : first aid, drinks, chocolate, an unsinkable knife, distress signals, flares, a florescent substance which leaves a green trail in the wake of the dinghies, a mirror, a floating anchor, a flag, a whistle, and a transmittor making location possible. All these items are light-weight and take up little space.

In the last War the main problem was water-supply. Two solutions were adopted : one chemical, developed by the British Admiralty and the other by evaporation, a system perfected in the USA.

The chemical method made use of mixed ions of silver and baryon which, with charcoal transformed the chloride and sulphate in sea water into salt. A filter sufficed to produce fresh water. The device in each dinghy measured 10 x 10 x 15 cms able to produce 5 times its volume of water, i.e. 2.5 litres. Several hundred thousands of these kits were issued to the RAF.

The American device was composed of a Solar Still containing sponges which, when soaked in sea water, mounted in a plastic balloon and being exposed to the sun, induced condensation and in fact fresh water that filled a container and cooled in contact with cold sea water.

A few life-boats came ashore on the coasts of Normandy and the French Resistance first hid their occupants, then arranged their passage along the escape routes to safety.

The RAF estimated that its Air-Sea Rescue Service along with the rapid intervention of R.N. Vedettes saved 3.306 crew-members almost one in two. Air-sea rescue today has profited from the advances mentioned above and added to them, notably the systems of detection-localisation by satellites : Argos, GPS...

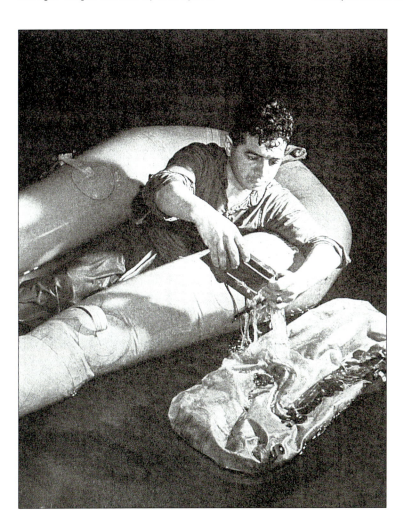

Inflated Life-raft fitted with Solar Still and condensation system to produce fresh-water

THE MIDGET SUBMARINE

Two Americans: David Bushell built the Turtle in 1776 during the War of Independence to torpedo English ships in the New York roads and Robert Fulton with the Nautilus in 1798, were the forerunners in the concept of the Midget Submarine. Robert Fulton came to offer his Nautilus to the "Directoire" in France, then at war with England. The little submarine 6 metres long and about 2 metres in diameter was displayed at Le Havre in 1800 and at Brest in 1801 and then in Paris on the Seine but failed to attract Napoleon's interest. John Philip Holland on 22nd May 1878 brought out the first midget submarine in the USA. The Czar too, produced a submarine, at first with pedals then in 1904 with an electric motor.

Not until the end of the First World War were midget submarines used in action, when, in 1918, the Italians attacked the Austrian Fleet in Trieste. Curiously enough, neither the Russians, the forerunners, the Americans, nor the French made use of the midget submarine in the Second World War. It was the Japanese, the Italians and above all the British who put flotillas of them into service.

It was these British "Midgets" Class X that would take soundings and bearings on the Calvados coast to draw up the marine charts for Operation Neptune. The tiny cabin: 1.6.m x 1.3m x 2.5m could squeeze in 3 to 5 men. This was true also of the X20 Exemplar and the X23 Xiphas that reconnoitred the Anglo-Canadian landing beaches. The X20 berthed submerged at Le Hamel to the west and the X23 at Ouistreham in the east from 4th June, awaiting the order for the Landings. On 6th June, still submerged they put out their goniometrical markings to guide in Allied shipping to their respective beaches. Had Operation Neptune been called off these two submarines could not have returned to England. They were to be scuttled and the crews to go ashore disguised as civilians with "real" false papers and ration cards. The Allies found with stupefaction, after the War that the Germans had a fleet of 50 midget submarines at Kiel copied from a British Midget captured at Kaafiord (Norway).

From X craft to Nautilus

This was the submarine of the Seehund type used by the French Navy until the end of the 50's. Midget submarines today however, no longer appear to be used by the world's navies. The knowledge gained in their development has been used to build craft for under water exploration. On 15th February 1954 Picard's Belgian FNRS remodelled by the CNRS descended to the Ocean bed to a depth of 4050 metres. Later on, both the American Alvin and the French Nautilus dived 6000 metres. It was the Nautilus from its "mother" ship the Nadir, after refinements by IFREMER, with a crew of three, that on 1st September 1985 discovered the wreck of the Titanic at a depth of 4 426 metres.

The Victor, the first tele-operated and guided non-crew submarine today, complements the Nautilus. It can operate 24 hours a day for several days.

British Midget Submarine X23 off-shore at Ouistreham

SURGERY

American Surgical Assistance Jeep in the Bocage, Normandy

500 years before our Era, Hippocrates the father of Medecine said: "If thou aspirest to become a surgeon, join the Army and follow it everywhere" Warfare is the making of medecine and in brief periods there is great progress to the benefit of human kind in general. New medecine: penicillin, sulphonamides, streptomycin...; blood transfusion and plasma; methods of anaesthesia and reanimation; new operating techniques, new surgical implements used by army surgeons have reduced by half, the deaths of the wounded in the two world wars

The paradox of Hippocrates

of the 20th Century. Surgeons operated close to the front line in the 600 hospitals and 40 000 beds had been installed in Normandy.

Anaesthesia was revolutionised entirely by the use of Penthotal which was used also to treat War Neurosis. Penthotal is also used notably for narcotic analysis and for "péridurales". Techniques of reanimation too have been made easier, notably with the use of oxygen. Treatment of burns by grafting skin has been made possible thanks to the use of dermatome, that makes it possible to take parts of healthy skin for grafting.

Aesthetic facial surgery has thus been able to limit the number of facial disfigurements so difficult to treat in the First World War. All the new techniques as well as those still unproven were in the hands of army surgeons.

To take account of the point surgery had reached at the time of the Landings, it should be mentioned that the surgeon Blalock in USA in 1944 had for the first time even operated on the malformation of the heart and had coined the term: "Blue".

Surgery in the front line in Normandy

DDT

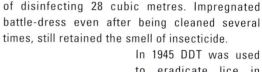

INVENTOR : P. Muller (N. P.)
DATE : 1939

DDT Dichloro-Diphenil-Trichloroethane was discovered in 1873 and used in dyeworks. In 1939, P. Müller and H. Martin working for Geigy, a Swiss company in Bâle, studying the effect of DDT on potatoes, discovered its value as an insecticide. It was a discovery of enormous importance that earned P. Müller the Nobel Prize in Medecine in 1948.

It was first patented in Switzerland in 1940, in Great Britain in 1942 and thirdly in USA in 1943. The use of DDT for an army was of great importance and an insecticide so cheap and easy to produce was termed the Saviour of Mankind. It was true literally on estimating that it saved 25 million lives across the world.

Military uniforms were impregnated with insecticide (as earlier in the War they had been against gas).

P. Muller

> **DDT
> the poisened
> chalice**

In Equatorial zones where malaria-carrying mosquitoes were prevalent; areas of water and swamp were sprayed from the air. In December 1943 and January 1944 a typhus epidemic in Naples was checked, in treating 1 300 000 Neapolitans.

In the 1944 Landings Allied troops had cartons of DDT powder in their kit. Aerosols were also issued capable of disinfecting 28 cubic metres. Impregnated battle-dress even after being cleaned several times, still retained the smell of insecticide.

In 1945 DDT was used to eradicate lice in concentration camps Unfortunately, the miracle insecticide had side effects on the environment largely due to its lasting effects. Traces of DDT were found in fish, birds and animals in places as far afield as Australia. It is estimated that of, say 3 million tonnes of DDT produced, half of it is still active. A million tonnes is still present in the sea, worldwide. In 1973 the industrial countries finally forbade the manufacture of DDT.

Eradiction of mosquito lavae by British troops

PENICILLIN

INVENTOR : A. Fleming (N.P.)
DATE : 1928

A. Fleming

for a cure for influenza, discovered "Penicillium rubrum" of the same family as "Penicillium album", present in the fermentation of certain cheeses. This substance arrested the proliferation of bacteria since it secreted an anti-microbian which is called penicillin. Since this was not yet stabilised and not what he was looking for, it did not become "the magic remedy" immediately; its use was for long delayed.

It was Howard Florey, professor at Oxford who, in 1938, succeeded in stabilising penicillin. Two years later B. Chain also at Oxford produced a purer substance, identifiable as an antibiotic that could be used clinically on mankind. Fleming, Florey and Chain all received the Nobel Prize for medecine in 1945. The appearance of penicillin complemented sulphonamides, by now seen to have limited uses and some side-effects.

The Leaven that saves lives

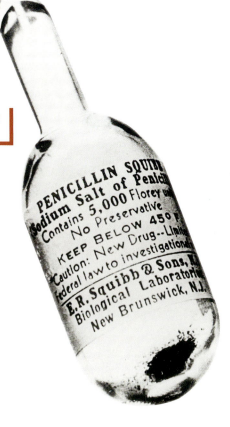

In a dispatch to the "London Gazette" of 3rd September 1946, Field Marshall Montgomery wrote: "Another interesting fact is that in the last War, of those with stomach wounds, two out of three died. Surgical units operating just behind the front line have now reduced this danger. In the Normandy campaign two out of three survived. The survival rate has been revolutionised by the use of penicillin". It is usual to ascribe this break-through to Alexander Fleming, British bacteriologist who discovered penicillin in 1928. Another Englishman Richard Barry, ascribes the discovery to a Frenchman Ernest Duchesne who had evidence in his research of the antibiotic qualities of "Penicillium Glaucun" in 1897 in Lyons.

Whoever it was and apart from this piece of research, Fleming, who like all the biologists of the time was looking

Thanks to PENICILLIN
...He Will Come Home !

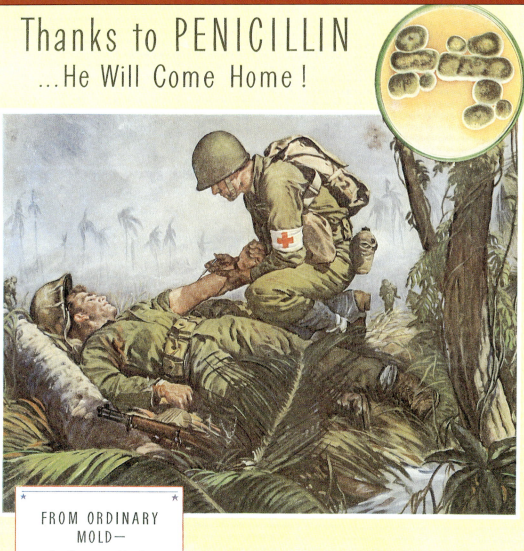

FROM ORDINARY MOLD—
the Greatest Healing Agent of this War!

On the gaudy, green-and-yellow mold above, called *Penicillium notatum* in the laboratory, grows the miraculous substance first discovered by Professor Alexander Fleming in 1928. Named penicillin by its discoverer, it is the most potent weapon ever developed against many of the deadliest infections known to man. Because research on molds was already a part of Schenley enterprise, Schenley Laboratories were well able to meet the problem of large-scale production of penicillin, when the great need for it arose.

When the thunderous battles of this war have subsided to pages of silent print in a history book, the greatest news event of World War II may well be the discovery and development — *not* of some vicious secret weapon that *destroys* — but of a weapon that *saves* lives. That weapon, of course, is penicillin.

Every day, penicillin is performing some unbelievable act of healing on some far battlefront. Thousands of men will return home who otherwise would not have had a chance. Better still, more and more of this precious drug is now available for civilian use ... to save the lives of patients of every age.

A year ago, production of penicillin was difficult, costly. Today, due to specially-devised methods of mass-production, in use by Schenley Laboratories, Inc. and the 20 other firms designated by the government to make penicillin, it is available in ever-increasing quantity, at progressively lower cost.

Listen to "THE DOCTOR FIGHTS" starring RAYMOND MASSEY. Tuesday evenings, C.B.S. See your paper for time and station.

SCHENLEY LABORATORIES, INC.
Lawrenceburg, Indiana

Producers of PENICILLIN-*Schenley*

STREPTOMYCIN

INVENTOR : S.A. Waksman (N.P.)
DATE : 1944

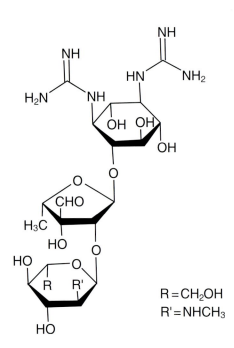

R = CH₂OH
R' = NHCH₃

R = CH_2OH
R' = $NHCH_3$

S. A. Waksman

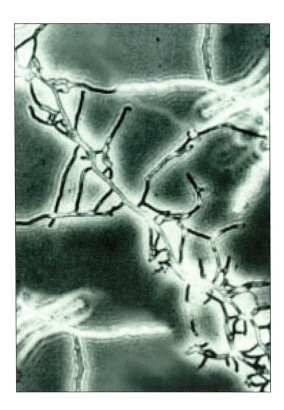

S. A. Waskman, microbiologist in the Department of Agriculture at Rutger's University, New Brunswick had since 1910 specialised in bacteria present in the soil. In 1940 he noticed that tuberculosis bacilli present in the soil were disappearing and concluded that other microbes, hostile were responsible. He invented therefore the term "antibiotic" : "a substance produced by a micro-organism capable of inhibiting or destroying other micro-organisms".

In 1942, amongst others engaged in research, he discovered streptomycin : particularly active against bacilli in tuberculosis and entrusted to one of his colleagues : A. Schatz the task of segregating streptomycin, achieved in 1943. S.A. Waskman was awarded the Nobel Prize in 1952.

Was streptomycin used as an antibiotic with sulpho-namides and penicillin by the American Army Medical Corps ? Some say it was used in Normandy in 1944 ; others aver it was not available until 1945. One thing is certain : information by Schatz in 1944 from which it may follow that some delay occurred for internal reasons in the USA. France however, in 1945, received 92 kilos of this precious medica-ment for hospital use.

SULPHONAMIDES

INVENTOR : P. Domagk (N.P.)
DATE : 1935

From 1932 to 1935 Paul Domagk, German Biochemist methodically studied thousands of chemical products to evaluate their anti-bacterial properties. He discovered that Protosil, a red pigment used for painting, first made in 1908, whilst being no use as it was, a slight change of formula rendered it most useful against streptococcus. He tested his findings on his daughter, suffering from a fatal infection of streptococci; she recovered. Domagk published his discovery and the

G. Domagk

medical profession found many uses for sulphonamide. In 1936 the remedy proved remarkably successful in checking an epidemic of meningitis that had broken out in the French Foreign Legion in Algeria. Paul Domagk was awarded the Nobel Prize for Medecine in 1939 but deprived of it by Hitler; he received it finally in 1947. During the D-Day Landings all the GI's carried a "First Aid Pouch" on their belts containing in a metal box with a red seal, a packet of sulphonamide powder.

BLOOD-TRANSFUSION

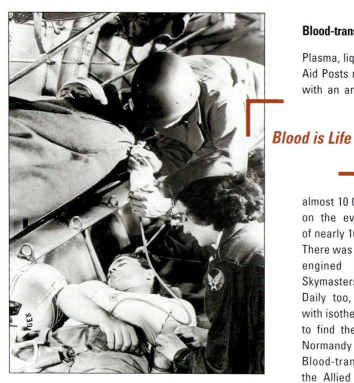

Blood is Life

Blood-transfusion in the front-line

Plasma, liquid or dehydrated was available in all First Aid Posts near the front line. Blood kept in condition with an anti-coagulant was stored in special flasks and maintained in refrigeration at 4 degrees. Collection of blood was particularly well organised. The USA could count on, up to 240 000 donors in one day. Blood banks in Britain and America had almost 10 000 litres each, of fresh blood in readiness on the eve of D-Day. There was a daily additio of nearly 1000 litres.

There was a daily liaison from the USA to Britain ; four engined Douglas freight planes, refrigerated Skymasters brought supplies for the Blood Banks. Daily too, fast light aircraft crossed the Channel with isotherme containers. Even now it is not unusual to find them in car-boot sales in towns along the Normandy coast. The logistics did not stop there. Blood-transfusion was of primary importance in the Allied Forces. Special convoys were escorted by Sherman tanks. Blood was even dropped by parachute when needed, in refrigerated containers, for it was summer in Normandy. Supplies of blood were, on occasion, even fired over the lines in shells, notably at Mortain.

At the beginning of the 20th Century an Austrian, Karl Landstreimer identified Blood Groups which reduced accidents in transfusions. He was awarded the Nobel Prize in 1930. If the number of accidents had diminished there were still some and in 1940 Landstreimer identified the Rhesus Factor which made it possible to complete the criteria of compatibility between donors and receivers.

From the outbreak of hostilities Britain and the USA created Blood Banks in which was stored blood and plasma in their natural state or dehydrated. Each had different uses. Plasma was used in the treatment of burns to re-establish the amount lost and maintain blood circulation, prior to hospitalisation.

Blood stored in isotherme flasks

A VACCINE AGAINST INFLUENZA

INVENTOR : J. Salk
DATE : 1933

Following the great epidemics of Spanish Flu in 1918 and 1919 that accounted for 20 and 40 million deaths more that is, than two world wars together, the Americans decided to look for a vaccine with which they could innoculate their GI's, prior to their return to Europe. During the great pandemic 800 000 of their soldiers had caught flu and 23 000 had died : more than they had lost in action in France at the time.

Of three Englishmen who worked on the research, W. M. Stanley identified a Flu virus first and was awarded the Nobel Prize in 1946. An American, Salk, in 1933, perfected the first anti-viral vaccine in 1933. Not until 1942 did American and Canadian research make possible the cultivation of the virus in the white of an egg.

In 1943 it was discovered that there were at least three types of virus : ABC each with a sub-virus A0, A1, A2, …and

The culture of virus in eggs

W. M. Stanley

that the spectrum of the vaccine had to be enlarged. As important as the vaccine's discovery was a method of mass-producing it. To do this on a grand scale presented a challenge. Millions and millions of fertile eggs were needed since the method consisted of placing a microscopic amount of virus in the air-pocket of each fertilised egg. Two to three days later about a table-spoonful of virus could be taken from each egg …

In this way from a batch of 150 000 eggs 1 000 litres of virus could be used to make the vaccine. An entire international network was put in place on the look-out for wherever the Flu virus could be found.

In an attempt to find an explanation for the outbreak of the 1918-1919 epidemic, the lungs were examined of a 22 year-old american soldier who had died 80 years earlier. The signs were of a virus of the type H1N1 similar to that of Swine fever, transmissible to Man. There were also exhumations of the bodies of miners in the north of Norway in the hope that the virus in the cold would lie dormant and could be identified.

VITAMINS

INVENTOR : F. G. Hopkins (N. P.)
DATE : 1906

A Vitamin Generator in Britain – August 1944

I n the course of his famous voyage round the World in 1740, Lord Ansen lost over half his crew who went down with Scurvy.

In 1770, Captain Cook found that his sailors recovered from scurvy if they ate citrus fruit (notably limes). Unconciously this famous captain had discovered vitamins. On his return to London, he sumitted a report to the Admiralty, noting his observation and on 7th March 1777 in a communication to the Royal Society he made mention of the useful role of citrus fruit.

Forty years later, the Admiralty included barrels of lime-juice in Naval rations. It took over 100 years to dispose of shipowners' objections and in 1894 the Board of Trade applied the same obligation to the British Merchant Service. The Army whilst acknowledging the need to eat citrus fruit, especially after the Crimean War, considered that it could not be expected to transport barrels of fruit juice to no purpose. Not until 1906 did Sir Frederick Gowland

F. G. Hopkins

Hopkins discover mysterious substances in food, termed : "accessory food factors" to which he gave the name of vitamins. He was awarded the Nobel Prize in 1929. In 1928 the Hungarian Szent Gyorgi discovered Ascorbic acid , Vitamin C in lemons. Haworth in England and Hirst at Reichstein in Switzerland both succeeded in producing a synthetic version in 1933. Artificial vitamins being absolutely iden-tical to the original, meant that the Army need not transport barrels; they could send out tablets, to provide vitamins for the troops.

Tobruk Tablets

The first tablets were despatched during the Siege of Tobruk in 1941 so that the garrison could survive and await reinforcements, thanks to these vitamin pastilles, which therefore came to be called : "Tobruk Tablets". The RAF too, issued their pilots with a cocktail of vitamines A and B to augment their sharpness of vision. At different times other vitamins were discovered : those soluable in water or fat : C and B... and the others : A D E ... The role of vitamins has been important enough for the Nobel Foundation to award prizes to no fewer than 8 scientists, involved in their development. In Occupied France when children suffered from malnutrition, "Marshal Pétain Biscuits" containing vitamins were distributed in schools. In Normandy, the Allies distributed vitamined chocolate to children. Later they did so in Paris and Marseilles followed. The Red Cross joined in by distributing vitamined jam to about a million other children.

JET AIRCRAFT

INVENTORS : von Chain and Heinkel
DATE : 1939

At the beginning of the 20th Century France held a dominant position in aeronautic development. "Aviators" from all over the world, flocked to France to exchange ideas, meet others, perfect and try out these machines that were heavier than air. Amongst them : Blériot, Wright, Farman, Santos Dumont, Roland Garros and Lübbe, innovators all, was a young French engineer : René Lorin who had thought up a new method of propulsion that would replace both the heavy thermal engine and the propeller. He described his jet-propelled engine in "L'Aérophile" magazine of 1st September 1908. He had thought out a new system of propulsion with an exhaust engine : the stato-reactor. He took out a patent for it in 1913.

In the First World War René Lorin wanted to build a Jet pilot-less plane, equipped with a gyroscope and a barometric altimeter to bomb enemy targets. He had therefore anticipated the V1 of the Second World War. It was too early and he was considered too much of an innovator; no one was interested in his projects. Dispirited, Lorin put all his theories in a book entitled : "L'Air et la Vitesse", published in 1930 by Editions Chiron. The book attracted mainly the Germans engaged in research !

The principle of jet-propulsion was not however, new. The Chinese and the Greeks knew about propulsion by gunpowder and by gas. Children the world over, know that a deflating balloon is propelled by the outrush of air. In 1928 F. von Opel in Germany flew, for several seconds, the first jet, rocket-propelled plane, in history. The same year an RAF officer cadet of 23, Frank Whittle, set out in his thesis a new system of Jet-propulsion capable of powering an aircraft. The device consisted of a compressor, an expansion chamber and a turbine. The era of Jet-propulsion by turbo-reactor had begun.

In January 1930 Whittle obtained his first patent and in 1936 he patented another one for a double flux reactor. That same year the Power Jet Company, set up with private capital the first turbo-reactor with a Whittle compressor that had "turned on the bench" on 12th April 1937.

From Roland Garros to Heinrich Lübbe

About the same time, the Germans, having started later, got there sooner. Actually von Chain, working with Heinkel also constructed a jet plane and for 8 minutes, flew the first turbo-reactor jet : the HE 178 on 27th August 1939. Germany remained ahead until the end of the War. During the 40's, Rolls Royce on the one side and on the other : BMW, Heinkel and Junkers developed a number of jet-propelled aircraft and equipped respectively: the Gloster Meteor, the Heinkel 162, the Messerschmitt 262, then the Arado 234 C the World's first four-engined jet, that flew first in February 1944. Every technological refinement had been attained, down to the ceramics on the Junker engines ! In 40 years Europe had invented, down to the last detail, the Jet-propelled Plane. Today, all the firms mentioned above, still flourish and are partners in the Airbus Consortium except one : Arado. In actual fact, its Chairman Heinrich Lübbe,

Arado 234 taking off in 1944

Heinkel 178 in 1939

a friend of Roland Garros in the early part of the Century, refused to belong to the Nazi Party and his firm was nationalised. It was wound up when the Regime collapsed. Now where is the moral in that?

To come back to Normandy in the summer of 1944: the Allied bridgehead was firmly established.

At the end of July the German GHQ sent out their system V reconnaissance aircraft: the invulnerable twin-engined Jet Arado 234 to photograph all the Allied positions. The mission took place on 2nd August Erich Sommer the pilot took his first sortie into hostile territory; he flew three times over Normandy and brought back dozens of remarkable photographs that laid bare the entire Allied logistics.

What all dreamed of doing Arado did !

TRANSPORT AND LOGISTICS

THE CONQUEST OF SPACE

On 13th June 1944 when the Allies were consolidating their bridge-head in Normandy, just one week after the Landings the first V1, fired at midnight fell on London.

A few hours later, on the same day, a V2 went out of control and came down in a marsh near Kalmar in Sweden. The country though neutral, promptly handed over what remained of it to the Allies. After a somewhat "rocambolesque" journey in a hearse it reached the coast, avoiding German attempts to recuperate it.

The era of missiles had begun; the V1 and V2 though very different in conception, led to the conquest of Space. The V1, put into service by the Luftwaffe, (German Airforce) was a small pilot-less plane. Catapulted from a reinforced concrete ramp, targeted on the objective, it was powered by an Argus Pulso-reactor. The idea had been taken from the book by René Lorin. The V1 flew at 600 km/h with a range of 250 km. Designed by Fieseler, 22,000 of them were built by Volkswagen at a cost each, of 600 US Dollars in current value. The V2 was a missile powered with Propergols, a mixture of oxygen and alcohol. It had a vertical take-off and was put into service by the German Army and fired by the Artillery. Like the V1 it was guided by a gyroscope and could not be put off course by counter electronic devices. It was largely supersonic and reached a height of 85 km. In weapons it was the absolute. First fired successfully on 3rd October 1942, ten thousand more were produced. The potential targets in the UK had to be within 250 km for the V1's and 350 km for the V2's. If, from the targets : London, Aldershot, Bristol, Southampton or Winchester one draws circles of 250 km and 350 km in diameter, the southern circumferences take in Northern Cotentin and the Seine Maritime. Considering the depots and factories of liquid oxygen it is understandable why Normandy played such a part in the German use of the V missiles. There were 65 V1 sites in the Cotentin and 116 in the Seine Maritime. It cannot be said with certainty where the sites of the V2's were, with the exception of a few large bunkers like Brécourt and depots like Hautmesnil,

A V1 about to be completed in an underground German Factory

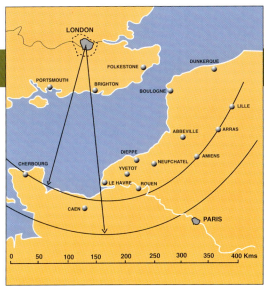

Rocket Sites
from which it
was possible to
reach London :
V1 : 250 km
V2 : 350 km

south of Caen, or the liquid gas stocks : other launch-pads were merely reinforced concrete ramps near a railway line, like the one at Le Molay Litry. It should be remembered that the development of these weapons cost the lives of some 20 000 deportees in the camps, Dora and Laura in underground factories that were never bombed, near Buchenwald. The work was directed by the Nazi, Wilhelm von Braun, using forced labour. He later became an American citizen ! After the War the Americans "recovered" 118 scientists, 250 V2's intact and hundreds of tons of material that was shipped to the USA from Cherbourg. Everything left on site was destroyed so that the Russians should find nothing. In spite of this, the latter gathered enough material to reconstitute almost 1000 V2's. How many scientists were taken to the USSR is not known. The Americans fired successfully, their first V2 in the desert of New Mexico on 3rd March 1946. But it was the Russians who came out best with these spoils of war. On 4th October 1957, to the stupefaction of the rest of the World, they launched the Sputnik.

The competition that followed is well known. On 20th July 1969 the Americans set foot on the Moon.

Europe took its time but with the Ariane "family", direct descendants of the V2, it succeeded in catching up, also with the help of German scientists. It may be noted that the propeller is produced at Vernon in Normandy.

From the cost in lives to the cost at the time in cash : Weaponry in US Dollars.

V1 V2	3000 million
RADAR (USA)	3000 million
Atom Bomb (Manhattan Project)	2000 million

From the Earth to the Moon

V2 preparing
for blast off

THE DAKOTA

Dakotas ready to tow gliders

Another example of early invention and late development is the DC3 or Dakota : a transport plane designed in the early thirties. An aircraft with many new features, it is still in service 70 years later. At the beginning of the Thirties travel by air was just starting but the aircraft of the time offered little in passenger comfort. Finding no suitable planes on the market TWA asked Douglas to design a new passenger plane capable of carrying in comfort 12 passengers.

In 1933, Donald Douglas, Kindelberger and Arthur Raymond brought out a plane with startling innovations. It had a metal frame reinforced by its metal casing; it was "fail safe" the DC1.

Only one plane was made but its remarkable qualities

A multi-purpose aircraft

appeared in a larger version, the DC2 and flew under the colours of TWA. Douglas assured them that the aircraft had a life of 75 000 hours, namely over ten times that of his competitors. History would prove him right.

Following the first commercial flights of the DC2, the famous DC3 was designed and the first test flight was on 17th December 1935 in Santa Monica. The Aircraft was put into service by the U.S. Army as the C47 and by the Royal Air Force as the Dakota. With a wing span of 29 metres and length of 19,5 metres, it had a range of 2 600 km. Originally it had twin-engines of 1 000 cv. It could carry 21, later 27 passengers or 4 tonnes

of freight at 340 km/h. Made entirely of metal it was the first mass-produced plane with an automatic gyro-pilot, Sperry giroscope. It cost $90 000.

During the Landings the DC3 was the general factotum: at dawn on 6th June hundreds of them dropped thousands of paratroops who, complete with equipment, were 18 to an aircraft and landed on the east and west flanks of the bridge-head. DC3's acted also as tug-planes for Waco-CG 4A (15 men) or Horsa (20 men) gliders. They were used later to bring in equipment and mail, taking wounded on the return trip. Some were used in precision drops and were fitted with "Mickey" radar to identify their objectives. A few were based at the air-field of Saint Aubin d'Arquenay from whence they towed back recuperated gliders to the UK and took part in the unfortunate Operation Market Garden on 17th September 1944.

Finally, they acted as transport planes ferrying goods to beseiged Berlin during the embargo imposed by the Soviets in 1948. Civil air lines took delivery of 800 DC3's whilst 10 928 were built for use by the Armed Forces, compared with 675 000 aircraft used over all, by beligerants of all the countries engaged in the 2nd World War.

A Dakota taking off with a glider in tow

THE WATER SUPPL

Reservoirs at Château de St-Gabriel

In 1900 Alexander a young Scots lad of nine, attending Clifton College Bristol took German and French, Latin and Greek. That is quite a lot for a youngster and Pindar's Ode to Nature went unlearnt. It was less pardonable however, since Alexander's class had recently visited Bath, a few miles away and engraved in the stone-work on the front of the pump-room of the Roman Baths was the first verse that he should have learned :

*Water, the Life-force for Victory ***

ARISTON MEN UDOR, WATER IS BEST, taken in the sense that from water, flows life itself. Alexander was given a hundred lines of that and never forgot it. He had a distinguished career : first as lieutenant in the Engineers in the First World War and later as a Civil Engineer. He returned to the Royal Engineers in 1939 and in the Landings of 1944 brought his Unit, 13th Airfield Construction Group ashore at Ver-sur-Mer. He was Colonel A.C. Rankin and his task was to construct temporary airfields where Allied aircraft could land from 11th June onwards. He was first at Crépon from whence he moved to Coulombs, destined to become the key Allied air-base. It was summer and despite the poor

weather, clouds of dust rose through the steel-netting runways when planes took off. A target for the Enemy, the dust spelt danger to the aero-engines too. The "Water Scheme" had been planned since January 1943 to spray the runways but drinking water for the troops was a priority. The lesson of the Lybian Campaign had been well learnt : "A battle may be lost for lack of a glass of water". The Water Scheme made provision for a double supply of water : drinkable and functional. First supplies were met from the UK in "Jerrycans", then army water-carts soon replaced by water from some twenty wells, sunk immediately following the landings on records supplied by geologists in Normandy.

In addition, water from neighbouring rivers and streams was drawn off, filtered and sterilised and supplied

Lt Colonel A.C. Rankin,
OBE, MC, DFC, Royal Engineers

"Water is Best ". An R.E. Pumping Station built at St Gabriel

by water-cart or as piped-water. From drinking-water for the troops we move to the second phase of the "Water Scheme"; a functional water supply with which Colonel Rankin had a primary concern.

About fifteen pumping stations were set up along water courses ; basins and sterilisation points were constructed and miles of water-pipes laid, through which passed millions of cubic feet of water to be sprayed every night on airfield runways to lay the dust. What now remains of this gigantic two-fold water-supply : enough to quench the thirst as well as float the Fleet ? A few wells are still in use, miles of underground piping with three pumping stations of a permanent nature still remain. One bears the inscription B.L.A. for British Liberation Army, another CARPE DIEM and the third ARISTON MEN UDOR. Alexander had learned his lesson !

This last station also pumped the water flowing from Fontaine Verrine in Saint Gabriel.

Colonel A.C. Rankin completed his tasks by constructing an airfield in record time, for Montgomery on entering Germany. The latter congratulated him and recommended him to H.M. King George VI for the OBE (Order of the British Empire).

*Agence de l'eau "Seine Normandie", Caen 1994

AUTOMATIC PILOT

INVENTOR : E.A. Sperry
DATE : 1912

To illustrate certain effects of the Earth's rotation as described by Newton 100 years before, Léo Foucault produced a device that rotated at speed which he described as a "gyroscope". The direction of the axis of rotation is unaffected by the movement of its base. It is this property that constitutes the unswerving direction of the gyroscope. In 1875 Henry Bessemer, British and Otto Schlick, a German used pig-iron as ballast in ships. At the beginning of the 20th Century the German Hermann Anschütz and the American Elmer A. Sperry used the gyroscopic principle to invent a compass for ships, the magnetic needle of which, could not be deflected by the electro-magnetic field created by the proximity of metal. They had produced the Gyro-compass. The use to which this invention was applied in 1912 to aeroplanes made the inventor famous. E.A. Sperry was an Engineer who had qualified at the American University of Cornell. His son Lawrence exhibited a Curtiss seaplane in 1914 in Paris, with an automatic pilot. He took off, put his arms above his head and his passenger standing on the wings. He created a sensation.

A Gyroscope Toy

And yet, it spins !

In 1917 Sperry fitted up gyroscopic torpedoes to be fired from aircraft and perfected the gyro-compass. In 1935 the DC3's were the first planes to be fitted with an automatic pilot. The course of the 2nd World War saw the advent of guided missiles, V1 and V2, guided bombs and torpedoes and many guided planes, one of which was the Arado 234. All these had gyroscopic equipment. Since then, many different types of gyroscopic guided aircraft equipment have emerged, in ever smaller and more sofisticated versions : laser driven and complementary to satellite navigation GPS (Global Positioning System).

Sperry Automatic Pilot

THE JEEP

INVENTOR : K. Prost
DATE : 1940

"The Jeep, the Dakota and the Landing-craft, were the three tools that won the War", said Eisenhower. Enzo Ferrari went as far as to say, "The Jeep was the only sports car that America ever produced" Almost mythical, symbol of freedom, for open spaces, rusticity, the Jeep was the forerunner of a long line of four-wheel drive vehicles for rough country, that manufacturers the world over put on the market in July 1940 , the U.S. Army sent out to some hundred firms a request to bring out

from ships.
Assembly workshops even, were set up in the Bocage so that a point in the Normandy countryside became an annex of Detroit.
On 14th June the first cases of component parts were unloaded at Arromanches and from 16th June the first Jeeps "made in Normandy" came off the assembly lines under the apple trees. Local productivity reached 100 Jeeps and 20 lorries daily.

When issued to them, the British nick-named the Jeep, the "pneumonia car" since it was so draughty, but it was in a Jeep that one of them, from 3rd. Infantry Division, was first into Caen on the morning of 9 th July 1944.

Almost a quarter of a million Jeeps were destroyed in the War and nearly 400 000 were sold in different countries.

It was adopted in Normandy for every part of work on the farm from towing farm machinery, to collecting milk cans... it is not unusual even now to meet one on the roads, in addition to those in the hands of connoisseurs.

**Jeep crossing Bretteville-l'Orgueilleuse
in the summer of 1944**

a reconnaissance vehicle suitable for rough country. A delay of 49 days with a maximum price of $175 000 was allowed to firms submitting offers.

Bantam replied, with Willys and Ford and was accepted at the figure of $171 185 and 75 Cents ! The Jeep (from G. P. General Purpose) was born, designed by K. Prost of Bantam.

*The Epitome
of the
4 W. D.*

Willys and Ford came in later and joined Bantam to produce the famous four-wheel drive. Mass production began from 1941 at the Willys Factory in Toledo and was to produce a vehicle worth $750 in the record time of 1 minute 20 seconds. The Jeep's baptism of fire was in Burma in 1942. 640 000 of these vehicles were produced during the War. During the Normandy Landings the first Jeeps came down in gliders, followed by thousands and thousands unloaded

**De Gaulle on the beach between Courseulles and Graye-sur-Mer
on 16 th June 1944**

THE LIBERTY SHIP

As German submarines sank more and more British ships in the Atlantic, Britain, in the emergency, ordered from the United States, still a neutral country, ocean-going merchant ships that could be mass-produced. On the strength of this "shipping order" and conscious too of the vulnerability of their own merchant fleet, a ship building project was put in hand for ocean-going cargo ships of an improved type called E.C.2, (Emergency Cargo ship 2) The Liberty Ship was born; it joined its mythical companions of the second World War, the jeep and the DC3. So it happened that the USA, long before Pearl Harbour, and for the first time anywhere, standardised ship-building for merchant ships at 10 800 tonnes, 135 metres long, equipped with engines of 2 500 h.p. with a speed of 11 knots. The E.C.2 provided for the building of 3,148 vessels from 1941 to 1945. There were 18 shipyards approved, 7 of which were controlled by J.H. Kaiser, champion of the new concepts of accelerated shipbuilding. In the various factories 19 000 different components were made and sent to the shipyards for assembly. Amongst other innovations, the Liberty ship was welded throughout. Each workman wore a belt with places for all the tools that he needed. At each work point, various types of gas were available : for welding drilling…

each on a bracket. A merchant ship was built in 30 days on average. The first Liberty ship was launched, unfinished on 27th September 1941. After Pearl Harbour on 7th December 1941, things speeded up. The race against time resulted in many imperfections. It was also decided to build a ship at such speed as to dissuade competitors. It was J.H. Kaiser's yard that met the challenge. Liberty ship No 440 christened Robert E. Peary was put on the stocks on 8th November 1942. She was launched on the 11th and sailed on the 15th. It had taken only four days, 15 hours and 25

A Liberty ship lying off Vierville

at Utah and four at Omaha. Germany artillery tried but failed to sink them prior to their arrival at the position planned. Amongst the hundreds of Liberty ships that plied between the Channel coasts, the "Jeremiah O'Brien" which made 12 crossings between Southampton and Utah Beach in the Summer of 1944, is the only vessel still afloat. She sails as a museum-ship. She made a trip to Cherbourg in 1994. France, as part of the lend-lease treaty, signed in 1946 received 75 Liberty ships, of which 30 were given names connected with the Battle of Normandy: *Argentan, Avranches, Bayeux, Bernières, Caen, Cherbourg, Courseulles, Coutances, Domfront, Falaise, Grandcamp, Granville, Isigny, Le Havre, Les Andelys, Lisieux, Mortain, Ouistreham, Pont-Audemer, Pont-L'Évêque, Port-en-Bessin, Rouen, Sainte-Mère-Église, Saint-Lô, Saint-Marcouf, Saint Valéry, Troarn, Valognes, Ville du Havre, Vire.*

4 days, 15 hours, 25 minutes who could do better ?

Launched by R.E. Peary
On 8th November 1942 at Richmond, USA

minutes to put her together. she sailed until June 1963 when she went for scrap. Kaiser had won hands down, though record building times varied considerably from yard to yard, some for example had no Slipway. There is an account of how one sponsor, arriving shortly before the launching, was amazed to see a bottle of champagne but no ship. He was told that; true she was a little late but would not be long! 2, 710 Liberty ships were built. Nearly 2000 were sunk. A few proved to have defects in their welding or showed a lack of buoyancy in comparison to ships with the traditional riveted hull. Plenty of Liberty ships were to be seen, off the Normandy coast from 6th June 1944 owards. Eight of them were towed in to be sunk to form the breakwater for Gooseberry 1 & 2, three

en 4 jours, 15 heures et 25 minutes

SUPERSTAR LOGISTICS

L ogistic : "The entire needs of a military force to conduct a prolonged action" , as Larousse has it.

To put in place the logistics of Operation Overlord was so considerable and complex that a few lines of explanation are necessary. The count-down began in April 1942 following the Boléro Conference in London when it was decided to cross the Channel and put an end to the German domination of Europe. A start was made and two years were needed to prepare the plan, train the troops needed and assemble material which had to be constructed and even invented and work out the lines of communication …

Eisenhower arrived in London in June 1942, to prepare the American contribution to the programme. It concerned the transport of 2 million men, 500 000 vehicles, millions of tons of food, petrol and ammunition !

Nearly 7000 ships were needed, most of which were purpose-built. They were to sail through five lanes marked out port and starboard and kept continuously swept of mines. Having obtained the relevant maps and charts, these had to be updated. In order to find convenient "hards" for landing, frogmen were sent out who recorded the declivity

12 tons for each GI

12 tons of equipment and supplies arrived with each GI

Beaches marked out for landing at low tide had to be prepared, using gear designed and purpose-built by the R.E.'s. A series of refrigerated containers to distribute blood for transfusions was put in place. Aircraft were ear-marked for rapid delivery of mail. Schemes for a network of oil and water-supply were developed. An overall plan of roads and streets was established and Units each received road-maps giving specific routes to avoid traffic jams. Where streets were too narrow, town bypasses were built.

Details of arrangements made, during and subsequent to the Landings were limitless.

D Day + 28 : The millionth man landed.

D Day + 38 : A million tonnes of goods and 300 000 vehicles had arrived.

D Day + 84 : End of August, 2 million men had landed in Normandy, 400 000 vehicles and 3 million tonnes of goods.

During this time : 2 prefabricated ports had been built, and 55 air-fields were laid out, the roads from Cherbourg to Caen and from Bayeux to Tilly had been widened. 1000 Bailey bridges were erected. A ring-road south of Caen, crossing the Orne at Athis farm was envisaged but not carried out until 50 years later ! What can be said of air traffic across the Channel and between the Normandy air-fields ? The flights amounted to tens of thousands and flight paths must have been adhered to scrupulously even during "round the clock" attacks on the Enemy, such as that of the afternoon of 7th August when Typhoons destroyed 300 German tanks at Mortain. Several hundred mobile radar units monitored Normandy

and took samples of sand on the beaches. From these findings nearly 200 million navigation charts were printed.

air-space for friend and foe alike. To all this was added the custody of POW's and the care of civilians and displaced persons.

BAILEY BRIDGES

INVENTOR : Donald Bailey
DATE : 1940

A Bailey Bridge in course
of construction near Bénouville
on the canal from Caen to the sea

About forty men were needed to assemble a Bailey Bridge; no mechanical gear was necessary. "Quite the best thing in that line we have ever had", was Montgomery's comment. It was a certain Donald Coleman Bailey of the British Ministry of Supply who, in 1940, brought out the pre-fabricated bridge that bears his name. Bailey Bridges are rather like a mecano set with identical parts (1,52m x 3,05m) 5 feet high by 10 feet long, easy to assemble by a few men. Each part consists of welded angle-irons that are bolted together to form a cantilever girder. These may be in parallel series of two or three relative to the weight carried. There were standard bridges of 57 metres for loads of 40 tonnes and 45 metres for 70 tonne loads. Since parts could always be added, the bridges were very versatile and could be put in use in anything between 14 to 34 hours. The bridges across the mouth of the Orne were 100 metres long. A few weeks later, bridges of 140 to 250 metres enabled the British to cross the Seine. Over 1000 Bailey Bridges were used in Normandy by the Allies. Practically all have now disappeared. Bailey or Panel or even Truss Bridges (Truss = Cantilever or Treillis) are still used in the American, British and Canadian Forces. They are often used in the event of natural catastrophes. They are also the object of competitions between units of the R.E.'s.

**40 men
to build
a bridge**

Churchill Bridge in Caen across the Orne

THE SCOOTER

The Larousse dictionary defines the Scooter as: a two-wheeled open vehicle where the driver sits astride. From 1936 onwards Cushman an American, designed vehicles that correspond to this definition and also tricycles with small wheels for the delivery of parcels, milk or even Coca-Cola. After USA entered the War, scooters were ordered from Cushman which, though not usual in Normandy, landed in Provence in August 1944 with US Airborne Troops. The "parascooter" later served as a liaison vehicle, notable for its Harley Davidson saddle, on US Army airfields in France and Italy.

In Italy, Enrico Piaggio was greatly attracted by this new type of cycle. He was the son of Rinaldo Piaggio who built vehicles of all kinds from aeroplanes to cars. It may be noted that on 22nd October 1939, the pilot Mario Pezzi flying a Caproni 161 with a Rinaldo Piaggio XI engine, in a 1.640 CV Canvas plane from Gnome and Rhône, beat the world record at an altitude of 17 083 metres, never surpassed by a piston-driven engine.

After the War, Enrico Piaggio, whose aircraft factory had been totally destroyed attempted to pick up the pieces. Cushman's scooter had doubtless given him ideas .

With what remained in his ruined factories, notably small aeroplane wheels, he produced a cheap and effective means of transport equally useful for work or pleasure to suit a large market in an Italy then devastated and poor.

At mid-day on 23rd April 1946 the Vespa was patented in Florence. A year later, Ferdinando Innocenti competing in the same market as Enrico Piaggio produced a similar cycle which he patented as the Lambretta. It is perhaps difficult to realise the almost mythical part they played in the social life of the youth of that period and today.

A Cushman Parascooter

PREFABRICATED PORTS

INVENTOR : H. Lorys Hughes
DATE : 1942

It was surely the most audacious technological adventure ever, but also the most controversial feature of the Landings. That is a maybe, but Mulberry worked and its remains are still to be seen offshore at Arromanches. Although the idea of bringing their port with them was first mooted in 1941 and the first study made began in 1942, it was not until August 1943 that the concept was finalised in the course of the Quebec Conference. During the voyage to Canada when Churchill and Mountbatten were on board the Queen Mary, a Mae West was thrown into a bath to demonstrate the effect of a floating break-water. Mountbatten brought his Staff Officers

into the bathroom for a demonstration. In the bath he put a fleet of ships made of newspaper which sank when the water was stirred up. He repeated the experiment with his little fleet surrounded by the life-jacket. The ships, protected from the swell, remained afloat. The concept of the prefabricated port, as a harbour safe for shipping had been proved valid.

The latter months of 1943 saw intense activity, working out the details of this ambitious project. The headquarters of the Engineer in Chief were in Kingswood School near Bath. His office was a room in the School known as the Governors' Room which looked out on to an old mulberry tree, hence the code-name of the prefabricated port. The challenging undertaking was as follows :

- There were to be two ports, one for the American sector to take 5 000 tons daily, the other for the British sector to take 7 000 tons of shipping daily.

- The 150 parts in steel or reinforced concrete were to be towed across the Channel.

- Each port was to take 10 Liberty ships at a time and the whole project was to be operational within two weeks of D Day.

- On 3rd September 1943 it was decided that each port should have an area of 600 hectares.

It was a gigantic task. This especially so for British economy and a nation that had already borne the weight of three years at war.

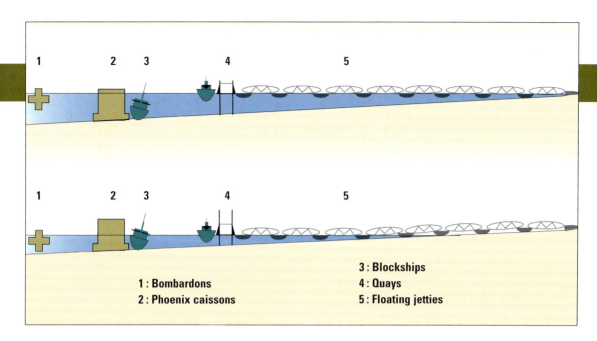

1 : Bombardons
2 : Phoenix caissons
3 : Blockships
4 : Quays
5 : Floating jetties

Views of the Port of Arromanches

Where would it be possible to find the welders, riveters, joiners and the millions of tons of steel and reinforced concrete, or the dry-docks already taken up with ships in for refitting; where could they get enough tugs?

Work only started in December 1943. Only six months remained to recruit 30 000 men and women and for the latter, their working week had to be reduced from 62 to 54 hours with no Sunday working. The first platform was launched on 26th January 1944.

The completed work: floating road-ways, pontoons, quays that went up and down with the tides break-waters, special anchors... were all ready, for the end of May 1944. It took in fact: 50 000 tons of steel and 600 000 tons of reinforced concrete, amounting to 3 million gross tonnage when the complete works crossed the Channel the day after 6th June. The challenge had been met. From 15th June onwards the first parts were in place. The storm of the 19th June destroyed the American port entirely; only Arromanches functioned normally. Built to take 7000 tons of shipping daily, rising sometimes to 12 000 the daily average was 6,765 tons. Port Winston, as it was called dealt with 2,5 million men, 500 000 vehicles and a million tons of goods. Useful or not? Prestigious or plain folly? Would beach landings have sufficed? These queries remain. It can be said however that the techniques learnt with floating quays have been used for the off-shore oil-rig platforms in use now.

The Story of Oil-rig Platforms

THE TURBOCOMPRESSOR

INVENTOR : Auguste Rateau
DATE : 1916

The first known air turbine was made in Alexandria 150 years before our era. It consisted of a large vertical tube which by the draught it created, spun a propeller on which revolved pictures from mythology. Not until last century with the development of aviation was any interest shown in air turbines or exhaust gases as a supercharge for engines. More than its known performance, it was its use in lack of air at high altitude that led to the use of the compressor in the First World War. It was a Frenchman Auguste Rateau who, in 1916 suggested the fitting of compressors as a supercharge. Two methods were used : to combine the compressor with the motor or to use the exhaust gases passed through a turbine. The first method : that of coupling the compressor to the motor was soon abandoned since at a certain altitude the power used by the compressor was greater than it produced. The second method, the turbocompressor, using energy from the exhaust gave greater force. Of all the aircraft fitted with a supercharge of this nature, two are worthy of note. They were engaged in a dog-fight at 44,000 feet, a record altitude. One was a Junkers 86 fitted with a German heavy oil two-stroke diesel and a compressor attached, piloted by Horst Götz, (of whom more later as pilot of the Arado 234). He was intercepted over Christchurch on his way to bomb Cardiff. His adversary in a Spitfire with a Rolls Royce carburator engine fitted with compressor was an Ace pilot of the RAF: Prince Emmanuel Galitzine, a descendant of Catherine the Great of Russia. They met at 44 000 feet in the first dog fight ever at that altitude, in a rarified atmosphere for men and motors alike.

**J.E. Johnson with his Spitfire IX
with a Turbocompressor**

The engagement lasted 45 minutes without either gaining advantage. Horst Götz returned to Caen-Carpiquet worn out. It was an engagement that remains without parallel in air warfare of the Second World War. After Mario Pezzi in 1939, they demonstrated that high altitudes were accessible to men and machines.

It may be noted in passing that Diesel aviation motors of the time were more powerful

*There is still air
at that altitude*

in relation to their weight than those fuelled by petrol. This perhaps explains why there is renewed interest today in diesel motors for light aircraft.

The turbocompressor is now used on many vehicles and recycles exhaust gases. It was first introduced by Saab for lorries in 1967, though it was Renault who won the French Grand Prix in 1979 with a Formula One driven by Jabouille, that put the turbo-compressor in vogue for cars.

WEATHER FORECASTING

In origin merely academic , the study of natural phenomena by Aristotal 350 years before our own times was summed up in his work, "Meteorology". It provided a starting point for what only became a science about the 17th century, with the invention of means of measuring : the thermometer by Galileo and the barometer by Torricelli. In 1892 two Frenchmen: G. Hermite and G. Besançon who, send up a balloon carrying a thermometer and a barometer took a step towards weather forecasts. It was with information gathered at sea, just after the First World War that aircraft obtained the data needed to cross the Atlantic.

In the Second World War, shipping and aviation pooled their resources. In fact mastery of the seas involved a knowledge of atmospheric conditions. The weather which might prove to be friend or foe, was increasingly better understood as the Conflict continued. It could be said for example that the weather befriended the Allies during the evacuation of the B.E.F. from Dunkirk in June 1940 with a clear sky and calm sea : that it was an enemy and befriended the Germans in June 1944.

Between the two dates however, great progress in weather forecasting had been made. In 1940 there were four flights daily to determine weather conditions for the Air Ministry; in 1944 there were thirty, some at an altitude of 44 000 feet. Use was also made of balloons and smoke shells and of course radar to locate cloud formations.

In 1942 the struggle to obtain mastery of the seas went hand in hand with assaults on weather stations. These, often camouflaged as fishing vessels made it possible to diffuse weather details gathered by ships. Up to that time radio silence was "de rigueur" for shipping, to avoid giving away their position to U boats. Morse by lamp was widely used , even in home ports.

On the eve of the Landings therefore, a five-day weather forecast was attempted. The weather, as is well known was deplorably uncertain in early June 1944 so that on the morning of 4th June, Operation Overlord was suspended. The Germans themselves considered any invasion at that time impossible.

Weather: both friend and foe

Eisenhower declared on that Sunday morning 4th June that the state of the weather was such that Allied air supremacy could not be guaranteed, that such was the key to success and that landing in Normandy must be postponed.

On the evening of the 4th at 21 hours Captain J.M. Stagg, in charge of weather reports, forecast a calm on Tuesday 6th though the final decision could not be taken until after a forecast had been made at 04.30 on the 5th that could confirm the forecast of the night before. The rest is History and Air Ministry Forecasts, along with other arms of the Service, had earned its reputation. It may be noted too, how weather forecasting can be a determining factor in the course of History. On 4th August 1944, a column of 400 German tanks, the armour from six divisions advanced on Mortain (on Hitler's orders of 2nd August). The aim was to cut off Patton's tanks from his base. The forecast was favourable ; low cloud gave cover from air attacks and success was assured. The forecast however allowed for possible cloud breaks about mid-day and Britain Typhoons « the tank busters » stood ready.

At 12.30 they began a "round the clock" attack that destroyed half the Enemy armour. Smoke from the burning tanks however was so intense that the attack was called off for nearly two hours , then continued until evening.

This was the first time that aircraft played a determining role in battle on the ground. Today, satellites involved in very sophisticated computer programmes have made possible considerable developments in weather forecasting.

Two members of the W.A.A.F. with a Theodolite about to take the bearings of a weather balloon

AEROSOLS

INVENTOR : F. Gauckard
DATE : 1939

Research on the use of gases in warfare, bacteriological or otherwise led to the development of moisture sprays containing chemical or biological micro-particles. It was a Frenchman, specialist in ultrafiltration, who invented aerolisation before the last War. In 1941 he brought to the Allies his plans and formula for aerosol bombs. The Americans produced an insecticide aerosol can, using a DDT base, filled first with carbon dioxide and later with Freon. It was used by them in the Philippines against the Anophele mosquito. During the Landings, all the Allied formations used aerosol insecticides in addition to powder. An aerosol can, could disinfect 28 cubic metres. Finally, regarding Japan, the Americans hesitated between the use of the atomic bomb and aerosol defoliants, sprayed from the air to destroy the rice crops; the resulting famine would supposedly have induced surrender.

As with DDT, Freon which is a threat to the Ozone layer was taken out of use.

THE RAY-BAN

INVENTORS : Bausch & Lomb
DATE : 1937

General Mark Clarck and General Eisenhower wearing Ray-Ban sun-glasses in 1943

After crossing the Atlantic with a balloon in 1921, (6 years before Lindberg's transatlantic flight), Lt John Mac Cready of the US Army Air Force suffered from headaches and nausea caused by the strong sunlight at high altitudes. A few years later the US Forces invited firms to bring out protective sunglasses with panoramic vision. It was a measure of protection for their pilots and an aid to spotting enemy planes in dive attacks out of the sun. Bausch & Lomb suggested a green tinted with an ultra violet and infra red filter and in 1936 put on the market an anti-reflection lens, accounting for one part of USAAF specifications. On 7th May 1937 Bausch & Lomb patented the name "Ray-Ban" (their product "banished" discomforting "rays") thus "Ray-Ban glasses with the large metal frales were born. They were "official issue" in the US Army Air Force but were widely used by all GI's."Ray-Ban" glasses landed on the Normandy coast in 1944. Apart from that, "Ray-Ban glasses" have, like the "Zippo", the "T-shirt", "Jeans" and other items become timeless and a tradition.

THE ZIPPO

INVENTOR : G. G. Blaisdell
DATE : 1934

The Zippo storm lighter, "the only lighter that never fails to light", said Eisenhower. It was on sale in all the PX stores (US equivalent of NAAFI) and landed in Normandy with the US troops. In company with the Ray Ban it became almost mythical.

In 1933, one: G.G. Blaisdell put together a lighter, current in the Austrian Army in the 19th Century and obtained a patent in 1934.

During the War he brought out two types of lighter. Until 1943 the metallic lighter in polished steel with an internal hinge in four sections was the only one on sale. It was succeeded by the "Black Crackle Finish" for reasons of economy and had a hinge of only three sections and was on sale to US Forces only. The black "crackle finish" paint, covering some base metal became the vehicle of expression as the equivalent of "Trenches Art" of the First World War. The GI's s cratched on the black paint their fantasies, achievements or merely a place or an event. The most worn or mostdecorated are particularly appreciated by collectors and considered of great value, especially as they carry a life guarantee, the maker undertaking to repair it without charge.

BALL-POINT PEN

INVENTOR : L. Biro
DATE : 1938

Although a patent had been taken out in 1888 by an American, J.J. Loud, for a large marker suitable for use on parcels, it was a Hungarian, Laszlo Biro who invented the ball-point pen in 1938. He had noticed that when children's marbles landed in a puddle, they left a trace on the ground. He patented his pen in Hungary before leaving for Argentina to escape the Nazis. There, he produced his pens under the trade mark : Birome. Just prior to the War he ceded the patent rights to Eversharp and its associate, Faber Inks. Reynolds also produced its own ball-point pen.

In 1943-44, the RAF having heard that this type of pen did not leak at altitudes, asked USA to supply them, with a pen of such magic qualities. It was to supply the air crews of bombers, exasperated by pens that leaked at varying altitudes and the indelible pencils of the time that left blue marks on the tongue. Quite often, still, at annual commemorations of the Landings, there are anecdotes of leaking pens and the welcome advent of Biro pens. Some veterans remember that back at base, the Biro made it possible to write in the rain.

The pen however was not without its problems : the shape of the point, the ink or not sufficiently, fluid, defective indication of ink level... Consumer complaints led Eversharp Reynolds and Douglas to abandon the American market to Parker.

In France, Baron Bich's tiny "Compagnie de Moulages" in Clichy also made pens; they approached Biro, obtained rights and put in a patent for "Carcasse pour un appareil scripteur" in 1951 and, in 1952, produced the "Bic Cristal". Striking success followed, due notably to an ambitious international policy. They took over Biro-Swan in England, and recently Schaefer and sell millions of Bic pens worldwide : a real success story – "à la française".

THE T-SHIRT

The T-type shirt or Training-shirt was a standard issue vest in the US Navy and the result of submissions by clothing firms in 1942. Contrary to what its name implies it was in fact an under garment worn for fatigues. Both Hanes and Union Underwear, two firms supplying the US Navy claim to have stocked it before 1942. The presence of American ships in all parts of the Globe assured a lightning success for the T-shirt. Like many other garments such as Jeans it became a tradition and basic for several generations. The T-shirt has now become a means of communication and its wearer a sandwich man.

The Sandwich Man

Base-ball players, US Navy

TECHNOLOGICAL ADVANCES
Savants, Inventors And Discoveries

850	KALDI	Coffee	Abyssinia	Page 8
1776	D. BUSHNELL	Submarine	USA	Page 34
1830	J. DEAN	Diving-suit	GB	Page 32
1838	A. SEIBE	Diving-helmet	Australie	Page 32
1839	C. GOODYEAR	Vulcanised rubber	USA	Page 11
1852	L. FOUCAULT	Gyroscope	Fr	Page 54
1858	NADAR	Aerial Photography	Fr	Page 22
1870	H. FLEUSS	Diving equipment	GB	Page 32
1891	G. LIPPMANN (N.P.)	Colour Photography	Fr	Page 24
1895	W. C. RÖNTGEN (N.P.)	X Ray	D	Page 28
1901	V. GRIGNARD (N.P.)	Silicones	Fr	Page 15
1906	F. G. HOPKINS (N.P.)	Vitamins	GB	Page 45
1908	R. LORIN	Stato-reactor	Fr	Page 46
1912	E. A. SPERRY	Gyroscopic guidance	USA	Page 54
1919	H. KOCH	Enigma coding machine	H	Page 16
1922	H. STANDINGER (N.P.)	Polymerisation	D	Page 13
1923	F. FISCHER - H. TROPSCH	Synthetic motor fuels	D	Page 12
1928	F. VON OPEL	Jet propulsion	D	Page 46
1928	A. FLEMING (N.P.)	Penicillin	GB	Page 38
1928	W. M. STANLEY (N.P.)	Influenza virus	GB	Page 43
1928	F. WHITTLE	Turbo-reactor	GB	Page 46
1928	P. PFLEUNEUR	Magnetic tape	D	Page 21
1928	H. T. KALMUS	Technicolor Process	USA	Page 24
1930	W. H. CAROTHERS	Nylon	USA	Page 13
1930	J. A. NIEUWLAND	Synthetic rubber	USA	Page 11
	W. H. CAROTHERS	Neoprene		
1931	MISHINA	Permanent magnets	J	Page 18
1933	J. SALK	Influenza vaccine	USA	Page 43
1935	AEG - TELEFUNKEN	Tape-recorder	D	Page 21
1935	G. DOMAGK (N.P.)	Sulphonamides	D	Page 41
1935	GERMAN CO-OPERATIVE	Buna synthetic rubber	D	Page 11
1936	A. TURING	Artificial Intelligence	GB	Page 17
1937	NESTLE	Nescafé	CH	Page 8
1938	L. BIRO	Ball-point pens	H	Page 67
1939	VON OHAIN - HEINKEL	Jet Aircraft	D	Page 46
1939	P. MULLER (N.P.)	DDT	CH	Page 37
1940	K. PROST	Jeep	USA	Page 55
1940	D. C. BAILEY	Bailey Bridge	GB	Page 60
1940	U.S. GOVERNMENT	Liberty Ships	USA	Page 56
1942	VON BRAUN	V 2	D	Page 49
1944	S. A. WAKSMAN (N.P.)	Streptomycin	USA	Page 40

(N.P.) : Nobel Prize winner

Acknowledgments

To all those who have assisted whether orally or in writing with their recollections, documents, even articles current at the time and not least for all the lavish encouragement :

Dr. J. P. Bénamou, M. Brissard, E. Chaunu, Colonel G. Legout, D. Mary (who sadly, will not have the chance to see this in print), Dr. J. P. Rioult, F. Robinard, General Scott-Bowden, J. M. Selles et E. Allain, E. Sommer, T.B. & M.M. Greenhalgh

Firms : Bic, Bausch & Lomb (Ray-Ban), Motorola, Thomson-CSF,

and especially to Admiral C. Brac de la Perrière, Mayor of Luc-sur-Mer and President of the D Day Commemoration Committee who so kindly wrote the Preface to this work.

Photographic credits

P. Bauduin, CAEN - DITE USIS, PARIS - Nobel Foundation, STOCKHOLM - Imperial War Museum, LONDON - King's College, CAMBRIDGE - Musée de l'Air et de l'Espace, LE BOURGET - Musée Mémorial de la Bataille de Normandie, BAYEUX - Musée D-Day Omaha, VIERVILLE - National Archives, WASHINGTON - Public Record Office, KEW - Submarine Museum, GOSPORT.

OREP

E D I T I O N S

15, rue de Largerie - 14480 Cully
Tél. (33) 02 31 08 31 08
Fax : (33) 02 31 08 31 09
info@orep-pub.com

Graphic Design : OREP
ISBN : 2-912925-25-8
Copyright OREP 2003
All Rights Reserved

Legal deposit : September 2000

Printed in France